Design of Self-Assembling Materials

Ivan Coluzza
Editor

Design of Self-Assembling Materials

 Springer

Editor
Ivan Coluzza
Computational Biophysics
CIC biomaGUNE
Parque Científico y Tecnológico de
Gipuzkoa
Donostia/San Sebastián, Gipuzkoa, Spain

ISBN 978-3-030-10079-7   ISBN 978-3-319-71578-0   (eBook)
https://doi.org/10.1007/978-3-319-71578-0

This Springer imprint is published by the registered company Springer International Publishing AG part
of Springer Nature.
The registered company address is: Gewerbestrasse 11, 6330 Cham, Switzerland

Foreword

Self-assembly is the process by which a substance, exclusively driven by noncovalent interactions, spontaneously develops into a particular long-lived conformation with a well-defined structure. Such structures may be highly inhomogeneous but still exceedingly differ from random amorphous materials which are typically dynamically arrested. In fact, by having an easily accessible ground state, self-assembling materials have the property of forming structures characterised by few defects, and they often can adapt to changes in the environment. Self-assembly is the method that most biological systems use to control the synthesis of complex structures. If properly understood, the fundamental principles used by all living organisms can be applied to design artificial materials with novel properties, such as active response to the environment, catalytic properties or precise three-dimensional connectivity which could extend the possibilities of modern electronics based on surface lithography. In this book, we have collected five essential topics, distributed in five dedicated chapters, to characterise the complexity and possible applications of the technology that exists to design self-assembling materials.

Across the chapters, we aimed at providing an overview of natural systems and novel artificial systems that can reliably self-assemble in predefined structures avoiding amorphous arrested (glassy) structures.

Biopolymers such as proteins and DNA will have a major representation in this book. Proteins are fundamental building blocks of living organisms and viruses. Their function is encoded in the same structural elements that compose them. This encoding is a unique property that stirs the interest of scientists from very different backgrounds. However, they have reached a high level of complexity (making the control step very difficult), and they are optimised for particular tasks which may not meet the general needs. While there are many different proteins with different structures and functions, they are all composed of chains of the same 21 fundamental chemical units called amino acids. Although only a few sequences will determine a stable ground state, the chemical heterogeneity of the alphabet ensures an enormous variety of combinations, each of which folds into complex arrangements of predominantly two types of secondary structures: alpha helices and beta sheets. Due to this complexity, the system represents a big theoretical challenge.

Large attention will also be devoted to the recent development in the fields of self-assembling in the colloidal systems, in particular of patchy colloids. Colloids are an optimal system, not only because their interactions are highly controllable, but their size often allows for tracking and characterisation in real time using confocal microscopy. Moreover, colloids and, in particular, patchy colloids have been more and more used as a model system for the biological systems, including studies devoted to the understanding of protein aggregation and protein crystallisation. This book is timely since chemists and material scientists are starting to gain control over the shapes and on the local properties of the colloidal particles. Hard cubes, tetrahedra, cones, rods as well as composed shapes of nano or microscopic size have made their appearance in the labs, and will hopefully become available in bulk quantities soon. Patterning of the surface properties of these particles is adding further directions to the anisotropy axis space. Patches on the particle surface can be functionalized with specific molecules (including DNA single strands) to create hydrophobic or hydrophilic areas, providing specificity to the particle-particle interaction. In the same way, as sterically stabilised colloids have become the ideal experimental model system for investigating the behaviour of hard spheres and simple liquids, the new physicochemical techniques will soon make available to the community colloidal analogues of several molecular systems. The overlap between biopolymers and colloidal science has been made more evident by the recent development as, with different techniques, chains and membranes of colloidal particles have been synthesised.

All the subjects as mentioned above are crucial for the understanding and control of designable self-assembling systems. However, scientists from such distant disciplines rarely have the chance to come together and exchange their point of view on the common problem of self-assembling. With this book, we intend to create a compendium that hopefully could be a reference for all that try to embark on the effort of the design of self-assembling systems both biological and artificial. Hence, in order to cover the broad spectrum of the subject listed above we have collected the contributions of scientist from the communities of polymer science, biophysics and colloidal sciences, both with a theoretical and experimental background.

Barbara Capone and Emanuele Locatelli have contributed with their extensive experience in the modelling of polymers solutions in Chap. 1 dedicated to "Design of Polymeric Self-assembling Materials and Nanocomposites in the Semi-dilute Density Regime: Multiscale Modeling". Barbara Capone is an expert in the coarse graining of polymers in the semidilute regime. She did her PhD at the University of Cambridge with Jean-Pierre Hansen, and after holding a position as a university assistant at the University of Vienna, she won an APART fellowship from the Austrian Academy of Sciences. Currently, she is working at the Università degli Studi Roma Tre with a Marie-Curie Intra-European Fellowship. Emanuele Locatelli is an expert in the statistical mechanics of system out of equilibrium and polymer simulations. He did his PhD at the University of Padova in Italy and is now a university assistant at the University of Vienna.

Emanuela Bianchi is the author of Chap. 2 "Modeling the Effective Interactions Between Heterogeneously Charged Colloids to Design Responsive Self-assembled Materials" dedicated to the recent advances in the modelling of colloidal systems. The chapter focuses on heterogeneously charged colloids and serves as an excuse to introduce the broad applications of colloids in material science. Emanuela Bianchi is a recognised expert in the field of colloidal solutions with an emphasis on the properties of functionalized particles also known as patchy colloids. She did her PhD at the University of Rome "La Sapienza" after which she won a Humboldt grant to work in Dusseldorf and a Lise Richter grant in Vienna. Currently, she is a university assistant at the University of Vienna.

Chapter 5 by Oriol Vilanova, Valentino Bianco and Giancarlo Franzese is entitled "Multi-scale Approach for Self-assembly and Protein Folding" and will introduce the other half of the sky in the self-assembling universe occupied by biopolymers.

Oriol Vilanova is finalising his PhD in Physics at the University of Barcelona focusing on modelling how proteins interact with nanoparticles in biological systems.

Valentino Bianco is an expert in the field of protein modelling in explicit solvent. He did his PhD at the University of Barcelona, and after a postdoc at the University of Vienna, he now holds a Lise Meitner fellowship in the same University.

Giancarlo Franzese is a world expert on anomalous properties of water and their influence on the folding and organisation of proteins. He is Professor of Physics at the University of Barcelona where he ranks among the top faculty members for research. He won several international prizes for his research, including the Royal Society of Chemistry-UK selection as Emerging Investigator in Soft Matter in 2012 and the ICREA Academia award in 2016.

Flavio Romano and Lorenzo Rovigatti authored Chap. 3 entitled "A Mesoscopic Computational Approach to DNA-Based Materials" that offers an extensive overview of the broad applications of DNA for the realisation of designable self-assembling materials. Flavio Romano and Lorenzo Rovigatti are both experts on functionalized colloidal systems and the development of coarse-grained models of DNA. Flavio Romano did his PhD at the University of Rome "La Sapienza", and after a postdoc at the University of Oxford now is a tenure track researcher at the University Ca' Foscari of Venice in Italy. Lorenzo Rovigatti did his PhD at the University of Rome "La Sapienza", after which he won a Lise Meitner fellowship at the University of Vienna and a Marie-Curie Intra-European fellowship at the University of Oxford. Currently, he is a postdoc at the University of Rome "La Sapienza".

The last chapter of this book is authored by Peter van Oostrum and is dedicated to "Experimental Study of Self-assembling Systems Characterised by Directional Interaction".

This chapter is crucial to put all the subject discussed from the perspective of possible experimental realisation and offers a set of examples of the fundamental techniques used to study experimentally self-assembling systems. Peter van Oostrum is an expert in the synthesis manipulation and imaging of colloidal solutions. He did his PhD at the University of Utrecht in the Netherlands, and he holds now a position as a university assistant at the University of Natural Resources and Life Sciences (BOKU) in Vienna.

San Sebastián, Spain

Ivan Coluzza

Contents

Design of Polymeric Self-Assembling Materials and Nanocomposites in the Semi-dilute Density Regime: Multiscale Modeling

1

Barbara Capone and Emanuele Locatelli

1.1 Introduction

Creating novel building blocks, which allow for an easy and large scale fabrication of complex materials, is a challenge and a central goal of diverse scientific fields, ranging from Physics to Materials Science. Over the past years, much effort has been devoted into creating tunable building blocks that could self-assemble and stabilize complex structures with particular features such as the diamond lattice, renowned for its important photonic properties [1]. This has motivated materials scientists to undertake a large scale analysis of building blocks of various shapes [2] and diversified functionalizations [3–5]. Colloids functionalized with attractive or repulsive regions (patches) [6–11] showed interesting self-assembly behavior, which can be tuned either changing by the shape of the patches or their relative orientation [12]. State-of-the-art methods for the synthesis of such particles include lithography [13, 14], microfluidics [15], or glancing angle deposition [16, 17]. However, shape-specific particle fabrication becomes more challenging when the particle has to be decorated with a large number of patches or if the shape and location of the patches has to be tuned and controlled with extremely high precision. Therefore, even though novel and interesting structures appear to be theoretically possible, they still remain hard to realize experimentally, given the precision required to achieve precise shape and relative orientation of the patches. Another extremely interesting class of functionalized particles is represented by

B. Capone (✉)
Dipartimento di Scienze, Università degli Studi "Roma Tre", Via della Vasca Navale 84, 00146, Roma, Italy
e-mail: barbara.capone@uniroma3.it

E. Locatelli
Faculty of Physics, University of Vienna, Boltzmanngasse 5, 1090 Vienna, Austria
e-mail: emanuele.locatelli@univie.ac.at

© Springer International Publishing AG, part of Springer Nature 2017
I. Coluzza (ed.), *Design of Self-Assembling Materials*,
https://doi.org/10.1007/978-3-319-71578-0_1

DNA coated colloids, where colloids are covered with a brush of double stranded DNA terminating into a short segment of single stranded DNA. The terminal part of the brush is functionalized to attach only to complementary DNA strands. Such a high selectivity in the hybridization should, on the one hand, allow for an extremely precise tuning of the final structure that the colloid is designed to self-assemble into but, on the other hand, it also induces a dynamical trapping into metastable structures. It has been shown that, in order to control and reduce the disordered region in the phase diagram of DNA coated colloids, it is necessary to re-introduce some entropy [18, 19]. In order to widen the region of the phase space in which the crystalline phase is either more stable or dynamically accessible with respect to the disordered one, it is therefore necessary to introduce polydispersity both in the interactions and in the length of the DNA strands [18, 19].

This chapter focuses on multiscale methodologies for polymers solutions, in particular on methods that allow to span from properties of single polymers up to properties of dilute to semi-dilute solution of polymeric systems, with different architectures and chemical details (Fig. 1.1).

A polymer is a large molecule (macromolecule) made up of many small chemical units, *monomers*, held together by chemical bonds or of *structural* items (spatially defined arrangement of two or more monomers) repeated regularly along the chain.

The chemical structure of a polymer is generated during the polymerization process, i.e. the process during which the elementary units are bound together via

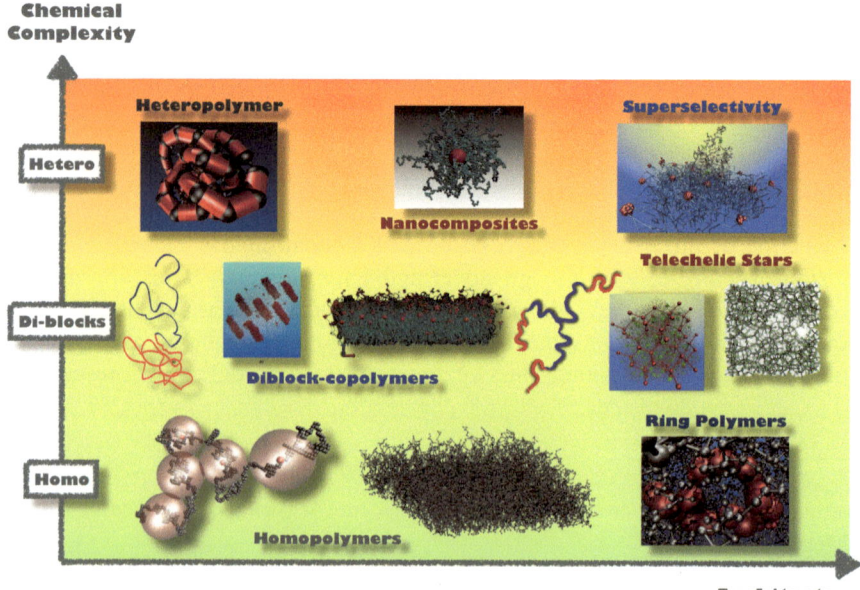

Fig. 1.1 A schematic guide to the eye on polymer solutions properties of changes in either chemical of physical complexity of the polymeric building blocks

covalent bonds. The number of units M in the sequence is called the *degree of polymerization*. M can vary: a short polymer has a degree of polymerization of the order of 10, while cellulose has $M \sim 10^2$–10^3 and polymers such as polystyrene can reach $M > 10^5$.

In polymer solutions the interplay between many degrees of freedom becomes important; the intrinsic nature of polymeric chains, the high number of monomeric units, and the dishomogeneity of distribution of polymeric segments lead to a complex framework where the contributions of many-body interactions play a crucial role.

The key to understanding complex fluids such as polymer solutions lies in the realization that it is possible to integrate out the degrees of freedom that are not of direct interest. The local microscopic features are important when looking for a specific practical application of a given polymeric material (e.g., for the optimization of the efficiency of a polymeric solar cell [20]). If, on the other hand, the goal of a study is to have a *global* view of the properties of classes of polymer solutions, it has been shown [21, 22] that omitting the chemical details of the chain structure as far as possible still allows to extract simple universal features of polymer chains.

The development of a multiscale coarse-graining procedure, that allows to reliably represent polymers at a given concentration, preserving the scaling and structural properties of the system, is an essential requirement to reliably predict properties of polymeric solutions, allowing for quantitative description, and mapping of theoretical predictions to experimental results.

1.2   Polymer Solutions

Physical properties of a polymer solution strongly depend on solvent quality (that can be affected and controlled by changes in temperature, pH), on the physical or chemical architecture of the macromolecules, and concentration of polymers in solution.

A broad distinction of solvents can be made dividing them in "good" or "bad" solvents; when polymeric chains are immersed in a "bad" solvent solution, monomers tend to minimize the contacts with the solvent, thus effectively giving birth to an effective intramolecular attractive interaction between monomers of the chain and intermolecular effective interactions between different chains carrying the same monomer–solvent interactions. The worse the solvent, the more compact the equilibrium structure of the chain.

Good solvents—on the other hand—dissolve polymers over a wide range of temperature; this leads to an effective interaction within the polymer (and between different chains) that is repulsive, since the polymer prefers to interact with the solvent rather than with itself (or with other chains).

As mentioned, the concentration of polymers also affects the properties of the system. Concentrations can be broadly divided into three different regimes: solutions might be dilute, semi-dilute, or concentrated (the latter is usually named *melt* regime).

Dilute solutions are such that each polymer does not feel the presence of the others, thus rendering a single chain properties analysis sufficient to describe the whole system.

In a *semi-dilute* solution polymers start to overlap, i.e. they are consistently in contact with each other: their distribution is locally inhomogeneous in space. The study of the properties of semi-dilute polymer solutions is of great interest; however, it is challenging both from an analytical point of view (we are faced with a spatially inhomogeneous N-body problem) and from a computational point of view because of the large number of microscopic units required for the estimation of thermodynamic quantities. It is then worthwhile to find a way to reduce the number of degrees of freedom of the system in a way that preserves the accuracy of results obtained microscopically for structure and thermodynamics, simplifying simulations and analytical calculations. We therefore introduce a so-called *coarse-graining* procedure.

We can finally introduce the third concentration regime: as the density increases the fluctuations become progressively smaller and the system can be treated by mean field theory. Such solutions are called *concentrated*.

In the following we focus on properties of uncharged polymers.

Different models have been introduced to describe, from a microscopic point of view, the effective interactions between uncharged polymers in good solvent. In the case of completely athermal solvents, simple lattice models such as the self-avoiding walk (SAW) can be employed; for continuous models, hard–sphere interaction between monomers is a common choice, although softer interactions such as a purely repulsive "truncated and shifted" Lennard-Jones potential have also been used. Extensive studies have shown [21, 23–26] that, in order to reproduce physical and thermodynamic properties and scaling laws, simple models are sufficient even though they ignore all chemical details, except for the excluded-volume effect and the polymer connectivity and architecture.

The simplest coarse-graining one could imagine consists of representing each polymer coil as an *effective sphere* [21, 22] of radius $r = R_g$, where R_g is the radius of gyration of the polymeric macromolecule, defined as the root mean square distance of each monomer from the center of mass (CM) of the polymer, centered in CM.

We can then introduce the *overlap density* as the density at which the polymers start to overlap: $\rho^* = 3/(4\pi R_g^3)$. For sufficiently long polymers (typically for $M > 100$), polymers are in the so-called *scaling limit* [27, 28] and it is possible to draw a relation between the radius of gyration R_g and the degree of polymerization M (i.e., the number of segments) of the polymer: $R_g \simeq bM^\nu$ where b is the segment length and ν is the Flory exponent which depends on the kind of solutions under consideration (i.e., for polymers in good solvent $\nu \simeq 0.588$, while for ideal polymers $\nu = 1/2$). It is then possible to write down a relation between the critical density and the length of a polymer chain: $\rho^* \simeq 3/(4\pi b^3 M^{3\nu})$. A semi-dilute solution is characterized by large and strongly correlated fluctuations in the segment density.

To define how each polymer interacts with the others in the coarse-grained description, we need to derive effective interactions between the new sub-units obtained by reducing the number of degrees of freedom.

The theory of effective interactions between polymer coils in the dilute regime started in the 1950, when Flory and Krigbaum showed that, within mean field approximation, the effective interaction between two self-avoiding polymers in a good solvent is Gaussian-like [22].

Over the past decade, considerable effort has been put in developing a simple but effective coarse-graining that could allow to describe the properties of dilute or semi-dilute polymer solutions. In [28], the authors introduced a systematic coarse-graining strategy which maps simple homopolymeric chains onto a fluid of "soft" particles interacting via effective pair potentials. For two polymers at infinite dilution, the effective interaction is computed by calculating the normalized probability $P(r)$ of finding the CMs of the two polymers at a distance r. The effective potential acting between the polymer is hence deduced as

$$\beta v(r) = - \ln[P(r)]. \tag{1.1}$$

For polymers in solution at finite concentration the effective potential $v_{\text{eff}}(\{\mathbf{r}_i\}, \rho)$ is a free energy which depends on the polymer density $\rho = N/V$ and on the configurations $\{\mathbf{r}_i\}$ of the polymers' CMs. A possible path towards the development of a coarse-grained strategy starting from zero density was developed in [28] where effective state-dependent pair potential $v_{\text{eff}}(r, \rho)$ at finite density ρ was extracted from the pair distribution function $g(r, \rho)$. It can, in fact, be proven [29] that for any given pair distribution function $g(r)$, at a given density ρ, there exists a *unique* corresponding pair potential $v_{\text{eff}}(r, \rho)$. The pair distribution function contains contributions that are of higher order than the pair potential $v(r, \rho = 0)$. It also contains all the many-body contributions arising from the interactions between polymers at a finite density. For this reason the pair potential $v_{\text{eff}}(r, \rho)$ is an effective potential and it is state (in this case density) dependent. Many strategies have been proposed to capture the many-body contribution on the effective potentials at finite densities as, for example, effective potentials can be extracted by inverting the $g(r)$ using the hypernetted chain (HNC) closure [28, 30]

$$g(r) = \exp(-\beta v(r) + g(r) - c(r) - 1) \tag{1.2}$$

of the Ornstein–Zernike equation.

Numerical results clearly show that the effective pair potentials computed at finite density present a nonnegligible density dependence. This is a clear demonstration that—as the polymer coils overlap—more than two-body interactions come into play, which result in a significant density dependence of the effective 2-body interactions [28]. The soft colloid approach on one side appears to be a simplification of the description of the polymer solutions, but on the other side it implies the need of computing a different effective potential for every density one wishes to analyze.

Following this approach requires that—in order to obtain a detailed coarse-grained description of a given polymer solution per every fixed state point—one needs to run full-monomer simulations at the state point in exam, extract the pair distribution functions between the CMs of the polymer chains and, subsequently, extract the state-dependent effective potentials, that can be then used to simulate the system on the large scale.

An alternative path is to shift away from a single-particle density-dependent description of the effective potentials to a modelization where density dependence is no longer embedded in the effective potentials, but within the coarse-graining.

In the following parts of this chapter a backtraceable multiscale methodology that allows to describe properties of polymers in the semi-dilute solutions is introduced. We first start with the simplest approach: the development of a multiscale strategy for linear homopolymers. We then add some topological complexity, describing ring polymers starting from the single molecule up to solution properties, to then change the chemical complexity developing a multiscale strategy for diblock copolymer solutions. Last we give an overview on the properties of a system that plays with both chemical and topological interplay: diblock copolymer star polymers, or telechelic star polymers, giving a glimpse on the self-assembling properties of what are shown to be molecular building blocks able to undergo a hierarchical steerable and tunable self-assembly process [31].

1.3   The First Step: Multiscale Representation of Homopolymer Solutions in the Semi-dilute Regime

The simplest system onto which such a representation can be applied is represented by homopolymeric solutions, and the alternative path cited above is to represent their properties in the semi-dilute polymer solutions by means of low density effective interaction potentials between two polymers.

In order to achieve this, we must reduce the finite density system to a *low density* case, and refine the coarse-grained representation as the density of polymers in solution increases.

The simplest system that can be described is made of N polymer chains, of M monomers each and length $L = (M - 1)b$, where b is the segment size. These N chains lie in a box of volume V, so that the density of the solution is $\rho = \frac{N}{V}$.

Many conformational, structural, and thermodynamic properties of semi-dilute polymer solutions, both in the bulk and under confinement, can be qualitatively understood in terms of scaling arguments based on the de Gennes–Pincus "blob" picture [21].

This picture has value whenever the characteristic length scale of the polymer solution (e.g., the correlation length ξ) or of the confinement is significantly shorter than the radius of gyration R_g of an isolated polymer chain. The blob picture suggests a systematic coarse-graining procedure whereby each polymer chain is divided into a number n of blobs, each containing the same number of monomers

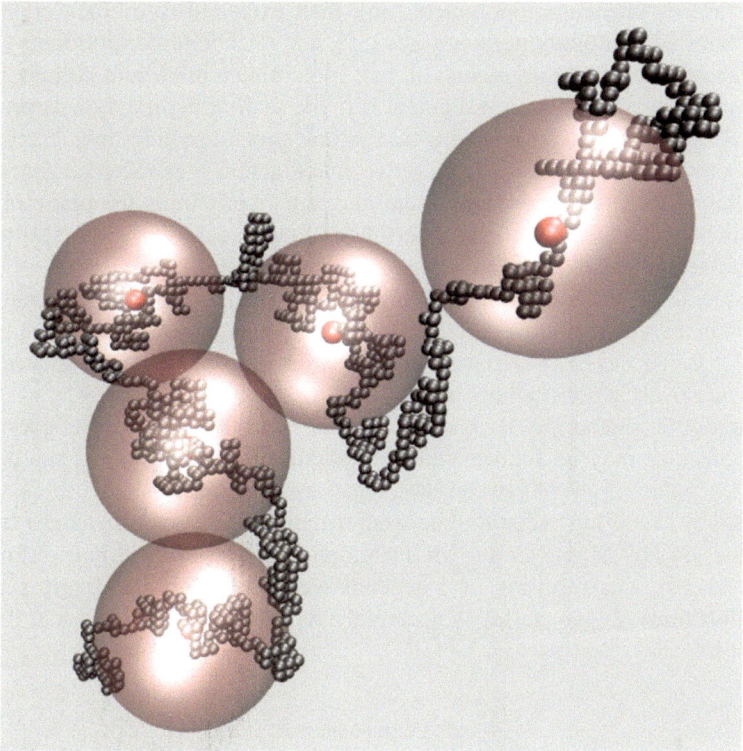

Fig. 1.2 A self avoiding polymer in good solvent made of 2000 monomers and its multi-blob representation

or segments $m = M/n$ of the initial coil, such that blobs of the same or different chains do not, on average, overlap, as sketched in Fig. 1.2 [32].

The multiscale representation that is introduced in what follows indeed consists in a chain-like description of each polymer: starting from a detailed full monomer description of the macromolecule, monomers are grouped in new sub-units. Each polymer chain of length L is thus divided into several sub-units, lowering the effective density of the system [32].

In this section we present and validate a quantitative formulation of such a multi-blob representation, which allows for a popular single blob coarse-graining procedure [22, 28, 33–35] to be extended to highly concentrated solutions. We then extend such a procedure to polymers and polymer solutions with different architecture (ring polymers, homopolymeric brushes).

A reliable coarse-graining, multi-blob methodology needs to be precise to be back-tractable onto the microscopic system; at the same time, it has to provide access to large scale analyses both in terms of the number of molecules in solution and in terms of the number of microscopic constituents. This section will describe a coarse-graining methodology as a first-principles approach, that was first introduced for the case of linear homopolymers [36], then extended to diblock copolymers [37,38], grafted homopolymeric brushes [39], and telechelic star polymers [31].

Chains are split into n segments (blobs), containing a minimum number of $m = 80$–100 monomers, so that each segment is in the scaling regime, i.e., its properties are not affected by finite size scaling and follow general scaling laws. Each one of the segments is represented via an effective potential that is extracted at zero density by simulating two chains, dividing them in dimers, computing the pair correlation functions between all dimers and inverting the analytical expansion that links the Mayer functions to the pair correlation functions [37,38,40].

Consider a solution of N self-avoiding polymer chains of $M - 1$ segments (each of size b) in a volume V; the polymer number density is $\rho = N/V$, and if $R_g \sim bM^\nu$ (with $\nu \simeq 0.588$ the Flory exponent) is the radius of gyration of an isolated chain, the polymer overlap density is $\rho^* = 3/\left(4\pi R_g^3\right)$.

In the dilute regime, $\rho^* < \rho$, where polymers do not, on average, overlap, the macromolecules may be represented by a single blob of radius R_g; the effective interaction potential $v(r)$ between the centers of mass (CM) of two blobs can be calculated by averaging over monomer conformations for a given distance r between their CMs, e.g. by Monte Carlo (MC) simulations of an isolated pair of polymers [22, 28, 33–35]. The resulting $v(r)$ depends weakly on polymer length L, and in the scaling limit $(L \to \infty)$, it is accurately represented by a Gaussian of width $\sim R_g$ [35]:

$$\frac{v(r)}{k_B T} \simeq A \exp\left[-\alpha (r/R_g)^2\right], \tag{1.3}$$

where $A \simeq 1.80$ and $\alpha \simeq 0.80$; the softness of the repulsive interaction, characterized by a modest free energy penalty of $\simeq 2k_B T$ at full overlap $(r = 0)$ of two polymers, reflects the low average monomer concentration $c \sim M^{1-3\nu} \sim M^{-0.77}$ inside each coil for long chains.

In the semi-dilute regime, $\rho > \rho^*$, polymer coils overlap, and this is reflected in a significant density dependence of the effective interaction [28], which spoils the simplicity of the coarse-graining procedure and introduces complications associated with state-dependent interactions [41]. This density dependence signals the fact that in the semi-dilute regime the relevant length scale is no longer R_g, but the shorter correlation length $\xi \sim R_g(\rho/\rho^*)^{-\gamma}$, with $\gamma = \nu/(3\nu - 1) \sim 0.77$ [21].

These shortcomings may be overcome by switching to a multi-blob representation, where the chain is composed by n blobs, each made up of $l = L/n$ segments. If $r_g \sim bl^\nu$ is the blob radius of gyration, the blob overlap concentration is $\rho_b^* = 3/\left(4\pi r_g^3\right) = \rho^* n^{3\nu}$. This means that the polymer density $\rho = \rho_b/n$ can be increased beyond ρ^*, up to $n^{3\nu-1}\rho^* \sim n^{0.77}\rho^*$ before the blobs overlap. In other words, the more blobs are used to represent one polymer, the deeper one can penetrate into the semi-dilute regime without significant blob overlap. Under those conditions the effective interactions between the CMs of the blobs are expected to be practically independent of blob density ρ_b, and may be taken equal to their zero density limit.

Blobs interact with one another via effective pair potentials that are extracted from simulations of two linear chains. In principle, one would have to consider two chains of n blobs each, and derive the effective potentials from those, by means of an inversion procedure. In practice, however, it turns out that considering two dimers consisting of two blobs each is sufficient.

Accordingly, the extraction of the potentials between the blobs results into coupled integral equations that include contributions arising from up to four bodies [36, 37]. The set of effective interactions consists in a Gaussian shaped potential that acts between all blobs (tethered and untethered) and a tethering harmonic potential acting between bonded blobs.

Such effective potentials are a universal property of the class of polymers analyzed, i.e., they are independent of the underlying model they have been extracted from [39]. In the blob representation, each of the n blobs has a radius of gyration, r_g; since within each blob there are m monomers, it holds $M = m \cdot n$. All the potentials acting between different blobs involve a single length scale, the radius of gyration of the blobs r_g. In particular, the blob–blob potential $V_b(r)$ acting between all non-neighboring blobs and the tethering potential $\varphi_t(r)$ acting between bonded, adjacent blobs, take the forms:

$$\beta V_b(r) = A \exp\left[-\gamma (r/r_g)^2\right] + B \exp\left[-\delta (r/r_g)^2\right] \qquad (1.4)$$

$$\varphi(r) = v(r) + \frac{k}{2}(r - r_0)^2 + c. \qquad (1.5)$$

These effective interactions thus include (or refer to) the pair potential $v(r)$ between non-bonded blobs on a given chain. The effective blob–blob interaction $V_b(r)$ is expected to be similar to the Gaussian repulsion in Eq. (1.3), with R_g replaced by r_g, i.e. the same as the effective potential between polymers in a single-blob representation [28, 34, 35]. $\varphi(r)$, on the other hand, may be expected to be the superposition of $v(r)$ at short distances r, and a harmonic spring at large elongation, similar to the entropic spring of a Gaussian chain [21, 25], albeit with a renormalized spring constant.

Once defined the set of potentials, each polymeric macromolecule is transformed into a chain of n-effective soft potentials, mounted to preserve the original architecture of the molecule.

The so-designed coarse-grained macromolecule satisfies all scaling laws of the original molecule; nevertheless the radius of gyration of the coarse-grained molecule differs—by means of a multiplicative constant—from the one obtained by means of a more detailed representation:

$$R_g = \alpha r_b M^\nu \qquad (1.6)$$

where αr_b is a prefactor that is linked to the model used, as it takes into account the exposure of the monomers to the solvent. Thus, to compare results obtained by means of a coarse-grained representation to a more refined one, or to an experimental realization of the system, it is important to define the relation between the radius of gyration of the microscopical system and the one measured using the multiscale method.

The radius of gyration of polymer follows Eq. (1.6), where ν is the scaling exponent, that—for example—for a polymer in good solvent is $\nu = 0.588$, for a polymer in θ solvent is $\nu = 0.5$, and for a polymer in bad solvent is $\nu = 0.3$, r_b is the bond length, characteristic of the system and α is a system-dependent prefactor, for example $\alpha = 0.44$ for a self-avoiding random walk or $\alpha = 1/\sqrt{6}$ for a polymer at θ solvent. For a general polymeric macromolecule α has to be determined, while the exponent is always the same given a specific state point (Fig. 1.3).

Now, by representing the polymer by means of a multiscale, multi-blob representation, the macromolecule does not follow the same statistics of the original chain any longer. If the original chain followed—for example—a self-avoiding statistics, the new coarse-grained chain made of soft-effective blobs does not follow a self-avoiding statistics. The radius of gyration R_g^b of a polymer made of soft blob should in principle follow the same relation in Eq. (1.6) but with a different prefactor $\alpha' r_b$

$$R_g^b = \alpha' r_b [r_g] r_g n^\nu = \alpha' r_b \left[\alpha \left(\frac{M}{n} \right)^\nu n^\nu \right] = \alpha' r_b R_g. \qquad (1.7)$$

Again, r_g is the radius of gyration of the blob, while $r_b [r_g]$ is the equilibrium distance of the blob tethering potential in units of r_g.

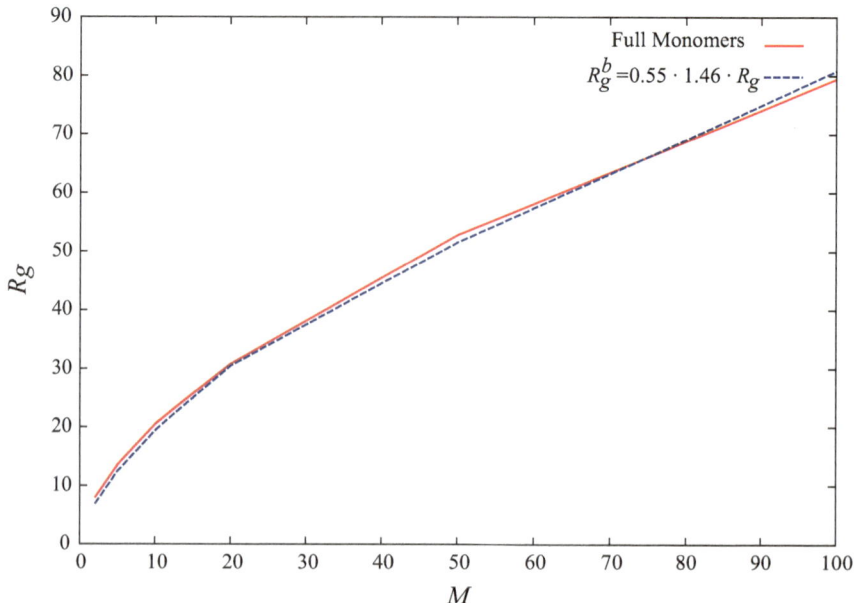

Fig. 1.3 Fit between the radius of gyration of a self-avoiding polymer made of blobs and one on a cubic lattice as a function of the polymer length. The length is in number of blobs and hundreds of monomers. The optimal fit is for $\alpha' = 0.55$

Simulations performed at various state points, from good solvent to theta and bad solvent, show that the radii of gyrations obtained by means of the multi-blob coarse graining follow the same power law in the number of monomers of the ones obtained by means of a more detailed representation. The coarse-grained methodology thus retains the scaling laws of the underlying microscopic models. Nevertheless, to fully compare multiscale coarse-grained results to the ones obtained either experimentally or with a full monomer representation, it is essential to rescale lengths with respect to the corresponding unit length of each specific model, e.g. the single molecule radius of gyration.

When different representations have to be compared, it becomes crucial to rescale all lengths by the respective radius of gyration computed at zero density, therefore obtaining results that are universal and independent of the description chosen.

1.4   Ring Polymers

Once the simplest multiscale strategy for homopolymeric solutions has been developed, it is possible to add topological constraints, i.e. molecules have particular architectures. Changes in topologies in the molecule affect both statical equilibrium properties, as the assembly behavior, and dynamical ones, as rheological response. A simple topological change on a homopolymeric molecule leads to the formation of ring polymers, i.e. macromolecules obtained by joining together the two free ends of a linear polymer chain. They are the most characteristic prototype of topologically constrained molecules [42], that allow to appreciate how the mere operation of "closing" a linear polymer chain has profound impact on the structural and dynamical properties of single molecules and concentrated solutions of the same alike.

Despite their conceptual simplicity and their highly interesting characteristics, the study of ring polymers, both experimentally and theoretically, is confronted with many obstacles [43]. From the theoretical point of view, the main difficulty of rings in comparison to their linear counterparts lies indeed in the treatment of the topological constraints, which, *inter alia*, prevents the formulation of the problem in terms of a field-theoretical approach [44] that has proven extremely fruitful for the treatment of solutions or melts of linear chains.

At the single-molecule level (equivalent to the infinite-dilution limit of a polymer solution), topology manifests itself in various ways. Although the infinite-dilution gyration radius of the rings, $R_{g,0}$, scales with monomer number M with the same, Flory exponent $\nu = 0.588$ as the linear chains in athermal solvents ($R_{g,0} \sim M^\nu$), topology effectively expresses itself as a larger excluded-volume parameter, resulting into a lowering of the Θ-temperature of the rings in comparison to linear polymers [45, 46]. A related, remarkable effect is the fact that in contrast to *ideal* (i.e., without excluded volume) linear polymers, ideal ring polymers experience an effective repulsion between molecules that is purely due to the additional topological constraint of closing each chain into a loop [47, 48], leading to a scaling $R_{g,0} \sim M^{\nu}$ that is identical to that of self-avoiding rings.

The effects of topology become even stronger at higher concentrations, and in particular at those exceeding the overlap density of the rings. Whereas the concentration screens out the excluded-volume interaction for linear chains, resulting into Gaussian statistics between the correlation blobs of the same [49], the topological potential between different rings cannot be screened out.

Solutions of ring polymers present a melt viscosity that is lower by one order of magnitude with respect to a solution of linear chains in the same density and solvent conditions [50–53]. Investigations on melts of unknotted, non-concatenated rings [43, 54–57] have shown that they display a higher diffusivity [43, 54–59] and that the Rouse regime extends to larger scales than in their linear counterparts [60]. Rheological experiments [61] and simulations [59] have revealed a power-law stress relaxation, instead of the usual reptation-like exponential behavior found for linear chains. Semiflexible rings, on the other hand, feature a particular form of self-organization in semi-dilute solutions, forming a disordered state of columnar clusters penetrated by other rings [48], and displaying an unusual dynamic scenario in which the coherent and the incoherent correlation functions are decoupled from one another, resulting into a state that has been termed *cluster glass* [62] (Fig. 1.4).

It is evident, thus, that the properties of topologically constrained molecules in the semi-dilute regime are extremely difficult to access, both via theoretical approaches and with computational studies. The difficulty of the latter increases with polymer size and density of polymers in solution, as the bigger the ring and the more rings in solution, the more the monomers to simulate, and the topological tests required in order to preserve the original topology of the system. It therefore

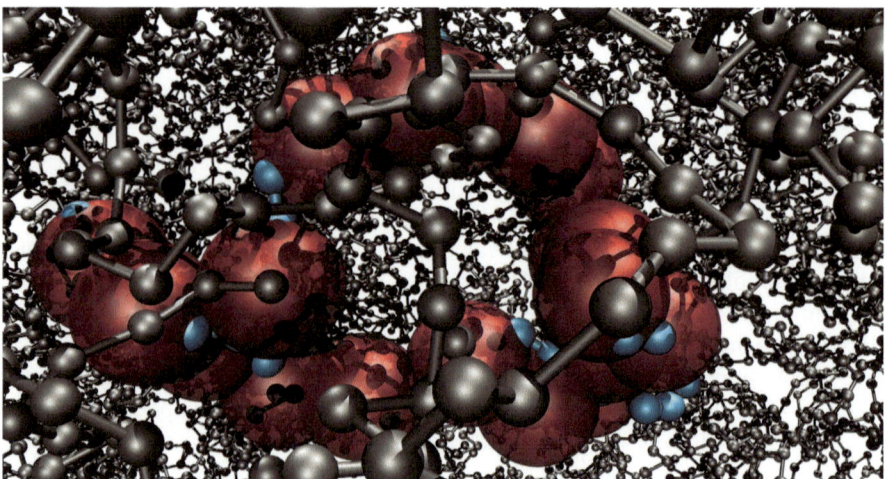

Fig. 1.4 A sketch of a coarse-grained polymer ring chain (red) in a solution of full monomer molecules (gray)

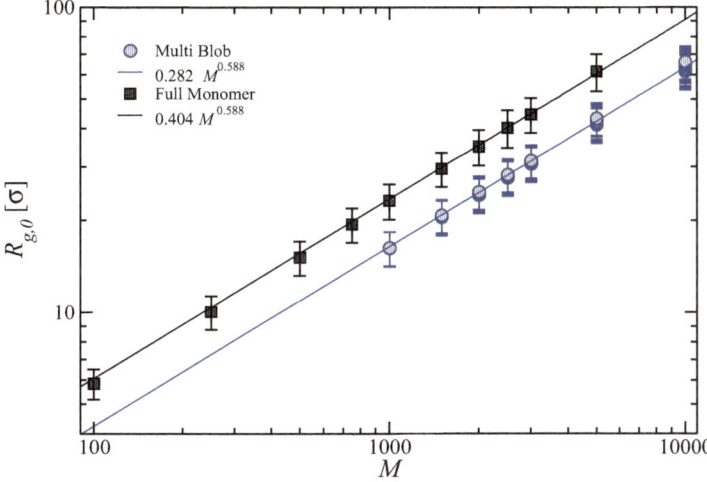

Fig. 1.5 Double-logarithmic plot of the dependence of the infinite-dilution gyration radius $R_{g,0}$ for flexible and unknotted ring polymers on the number M of microscopic monomers. Black squares correspond to full-monomer simulations and blue circles to the multi-blob model. Within the latter, different coarse-graining realizations resulting from various combinations of m and n at fixed values of $M = m \cdot n$ are shown. Dashed lines are fits to a power law with the Flory exponent $\nu = 0.588$

becomes of crucial importance to be able to analyze, simulate, and access the semi-dilute regime for molecules with a ring architecture.

As for the homopolymeric chains, it is important to define the ratio between the radius of gyration in the two different representations, in order to reliably switch between a multiscale representation and a more detailed one. Notwithstanding the topological constraint, the ratio between the radii obtained with the two representation shows the same scaling law and a different prefactor (see Fig. 1.5), that is preserved upon increasing density. Scaling all lengths with respect to the zero-density single-molecule model dependent radius of gyration allows to obtain model independent results, thus exploiting the potentiality of a reliable multiscale methodology.

Once the multi-blob procedure is set up, it is necessary to test the level of coarse graining necessary to represent a ring polymer without affecting its properties. In order to do this, single molecule properties such as shape parameters and the distribution of the radius of gyration are computed both with the coarse-grained representation and the microscopic one, and compared to one another. The same is done for the center-of-mass effective interaction. The knowledge gained from these comparisons also serves as a basis for deciding which is the minimal number of blobs per ring necessary for a faithful representation of the properties of concentrated systems.

We first consider single molecule properties. In order to assess the validity of the coarse graining procedure, we consider two characteristic quantities, namely

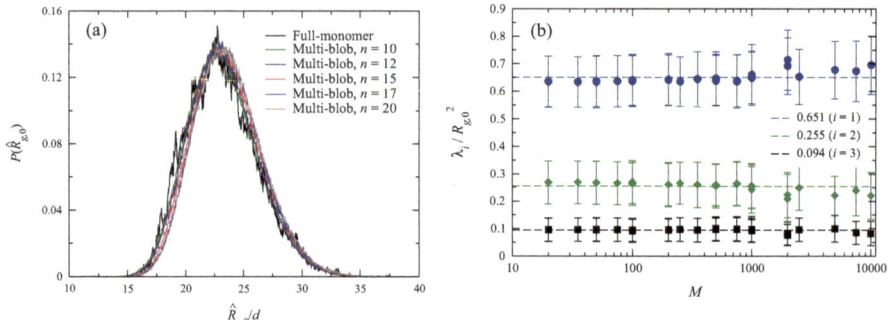

Fig. 1.6 The distribution functions of the instantaneous radius of gyration $\hat{R}_{g,0}$ of ring polymers at infinite dilution, obtained within full-monomer simulations and in the multi-blob representation at different levels of coarse graining. (**a**) $M = 1000$ monomers; (**b**) the three eigenvalues of the gyration tensor of flexible ring polymers, normalized to $R_{g,0}^2$, where $R_{g,0}^2 = \sum_{i=1,2,3} \lambda_i$, obtained both with a multi-blob representation and within full-monomer simulations; same symbols are used in the plot for both representations. Results are shown for single molecules of different lengths and different blobbing realizations for a given length. As all molecules presented in this work are in the scaling regime, the average shape of the single polymeric ring is influenced neither by the coarse graining level nor by the number of monomers in the macromolecule

the gyration tensor, which entails information on the shape of the rings, as well as the probability density $P(\hat{R}_{g,0})$ of the instantaneous gyration radius $\hat{R}_{g,0}$ of the rings. Both quantities have been measured within full monomer simulations and within the multi-blob approach for vastly different values of M and blobbing fractions, with the number of blobs per ring, n, employed in the latter lying in the range $10 \leq n \leq 20$.

The striking agreement obtained between the two representations for all M corroborates the validity of the multi-blob approach for isolated molecules and it establishes that a number of blobs as small as $n = 10$ is already sufficient to reproduce the salient features of ring polymers. At the same time, as each blob contains at least $m = 100$ monomers, a very fast simulation of a 10-blob-ring brings forward the properties of ring polymers with at least $M = 1000$ true monomers (Fig. 1.6).

Once ensured that both single molecule properties and effective pair potentials are reproduced by our coarse-graining methodology, we focus our attention to properties of semi-dilute ring polymer solutions upon augmenting densities. To deeply explore the semi-dilute regime, we performed with both representations Monte Carlo simulations of the full monomer system and of the multi-blobbed rings for a range of densities $0.5 \leq \rho/\rho_* \leq 7$.

Results from our simulations for the density-dependent radius of gyration R_g are shown in Fig. 1.7. The observed shrinking of the polymer size for $\rho/\rho_* > 1$ can be understood in the framework of the correlation-blob-model. The polymer consists of n_ξ correlation blobs, each of size ξ, performing a random walk characterized by some exponent υ. Accordingly, the gyration radius of the polymer depends on

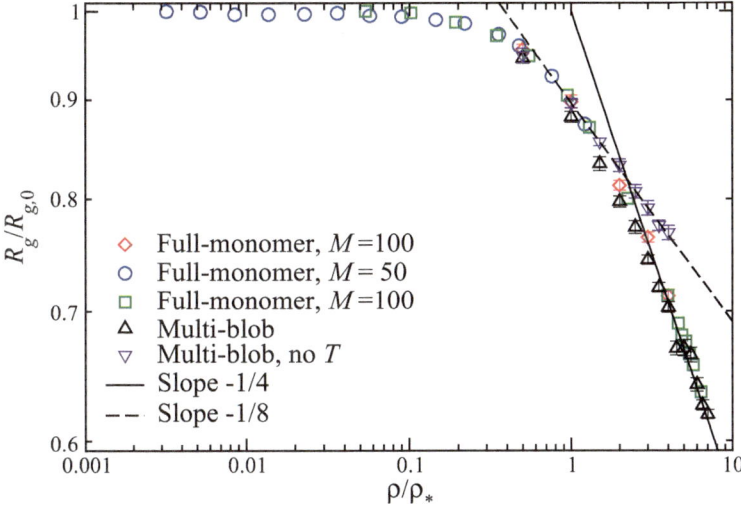

Fig. 1.7 The concentration-dependent gyration radius R_g of unknotted and non-concatenated flexible ring polymers, reduced over its value at infinite dilution, $R_{g,0}$, as a function of the ratio ρ/ρ_*. Results obtained from simulations of different models are shown, as indicated in the legend. The multi-blob simulations were carried out using rings of $n = 50$ blobs. For comparison, runs in which concatenations were artificially allowed were also performed, and the results from those are denoted as "no T". The two lines with slopes $-1/4$ (solid) and $-1/8$ (dashed) are shown to compare with theoretical predictions in which the topological interaction is taken into account or omitted, respectively

ξ and n_ξ as:

$$R_g \sim \xi n_\xi^\upsilon. \tag{1.8}$$

Using the definition of density and the scaling of the blob size, Eq. (1.8) yields a power-law dependence on density, namely

$$R_g \sim R_{g,0} \left(\frac{\rho}{\rho_*} \right)^x \qquad (\rho/\rho_* > 1), \tag{1.9}$$

with the exponent

$$x = -\frac{\nu - \upsilon}{3\nu - 1}. \tag{1.10}$$

Note that in the melt, where $\xi \to \ell$, where ℓ is the monomer size, and $n_\xi \to M$, Eq. (1.8) is valid for the polymer size at all scales.

If the correlation blobs also performed a self-avoiding random walk, we would have $\upsilon = \nu$ and polymers would not shrink at all for $\rho > \rho_*$. However, concentration screens out the excluded-volume interactions [49], and linear chains adopt in concentrated solutions Gaussian-walk conformations at scales larger than ξ. This implies $\upsilon = 1/2$, leading via Eq. (1.10) to the well-known result $x = -1/8$ for linear polymers [49]. For rings, things are different. The topological interactions between different rings cannot be screened out, thus the exponent υ cannot be the one corresponding to a Gaussian random walk. The value of this exponent has been an issue of a long debate. Early theoretical arguments [63], supported thereafter by computer simulations [43,54,55,57,64–66] seemed to converge to a value $\upsilon = 2/5$ ($x = -1/8$) for this exponent. However, more recent simulations with longer chains as well as more sophisticated theoretical approaches [44, 59, 60, 65, 67–71] show that the situation is a bit more subtle: For very long chains whose length M exceeds the entanglement length [72] M_{e}, the exponent is $\upsilon = 1/3$, corresponding to a collapsed lattice animal. However, for $M < M_{\mathrm{e}}$, a broad crossover regime that can cover several decades in M exists, and for which a power-law dependence with the exponent $\upsilon = 2/5$ is valid.

It is then possible to focus on the pair distribution function $g(R)$ between the rings' centers of mass, comparing the results from full-monomer simulations and from the multi-blob representation in Fig. 1.8. The excellent and parameter-free agreement between the two underlines the ability of the multi-blob approach to reproduce the structural correlations in concentrated ring polymer solutions. The clustering artifacts caused by the single-blob representation on the basis of $V_{\mathrm{eff}}(R)$

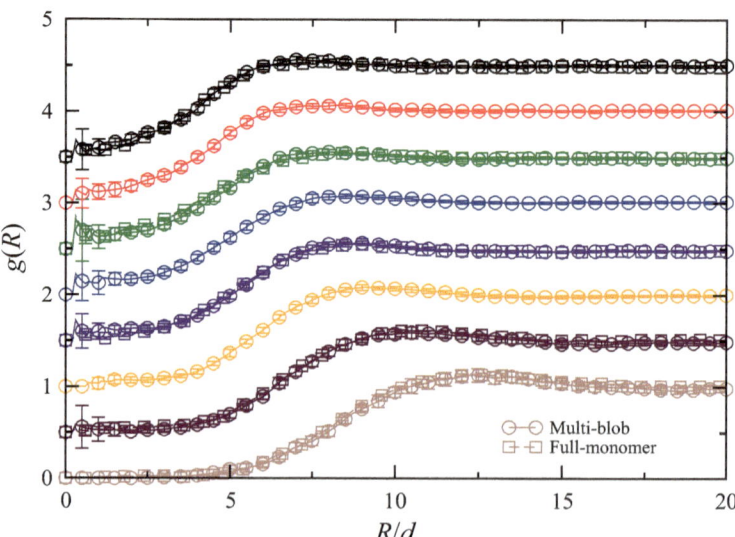

Fig. 1.8 The pair distribution functions $g(R)$ between the centers of mass of unknotted, flexible and non-concatenated ring polymers at various polymer densities ρ/ρ_*. Shown are results from both the full-monomer simulations with $M = 100$ hard monomers per ring (for $\rho/\rho_* = 0.5, 1.0, 2.0, 3.0$ and 4.0) and the multi-blob simulation with $n = 50$ blobs per ring. From bottom to top, densities are increased in steps of $\Delta\rho = 0.5\rho_*$, starting from the value $\rho = 0.5\rho_*$. For clarity, each curve has been shifted up by an amount of 0.5 from the preceding one

alone are removed [73]. No inversion procedure of $g(R)$ is any more necessary to yield strongly density-dependent, single-blob effective potentials [73]. Indeed, the latter are merely an expression of the increasingly strong many-body interactions that are obtained when one insists at describing mutually overlapping ring polymers as pairwise-interacting single blobs. Once the rings are divided into a sufficient number of smaller segments, each one still large enough to contain a large number of monomers, and these segments have overwhelmingly pair-contacts, the need to resort to density-dependent effective interactions is not present any more. The system can be described by means of fully transferrable, realistic, and reliable pair interactions alone.

1.5   Multiscale Coarse-Graining of Diblock Copolymer Solutions

Self-assembly and microphase separation of large molecules with dual physical or chemical functionalities are ubiquitous phenomena in complex fluids or soft matter, with far-reaching consequences in materials science or molecular and cell biology. One of the best-known examples is the formation of micelles, vesicles, micro-emulsions, and bi-layer membranes in solutions of amphiphilic or surfactant molecules formed of a hydrophilic polar head-group and one or several hydrophobic tails [74]. Similarly AB diblock copolymers in solution self-assemble into a wide variety of microphases, including lamellar, hexagonal, cubic, or bicontinuous phases [75]. While microphase separation in copolymer melts is well understood within self-consistent mean field theory [76, 77] and its improvements [78, 79], the theoretical understanding of self-assembly in copolymer solutions is much less advanced [80], mainly because polymers in solution no longer obey ideal (gaussian) statistics, and because the osmotic compressibility can be large, as opposed to the usual incompressibility assumption for melts. Monte Carlo (MC) simulations of copolymer solutions have been for long restricted to short copolymer chains [81–83].

Self-assembly and microphase separation generally arise from a competition between entropic (which oppose self-assembly) and enthalpic (which favor self-assembly) contributions to the total free energy. This competition, controlled by the temperature and the polymer concentration, leads to the very rich phase diagrams observed for various copolymers in selective solvent [75, 84]. For instance, the complex phase diagrams of styrene-isoprene diblock copolymers in various solvents have been mapped out experimentally in great detail by varying temperature, concentration, solvent selectivity as well as the relative sizes of the two blocks by a combination of light scattering, small angle neutron and X-ray diffraction and rheological studies [85, 86].

By analogy with the coarse-graining method used for homopolymer solutions, the main idea of the multi-blob approach for diblock copolymers is to eliminate any density dependence from the effective potentials used to represent the system.

Instead of using only two blobs, as was done in the dumbbell description, we need to increase the number of blobs used to represent the polymer. To do this we have to correctly quantify the number of blobs needed to represent the polymers for each density of the copolymer solution. As shown in Sect. 1.3 for homopolymers, we use the definition of the overlap density as a function of the radius of gyration R_g of the chain to determine the number of blobs. We then define the radius of gyration r_g of each blob as a function of the total number of monomers M of the polymer and of the number n of segments we are dividing the chain into:

$$r_g \sim b(M/n)^{\nu}, \tag{1.11}$$

ν being the Flory exponent, and b a factor linked to the solvent quality and the bond length between subsequent monomers.

The main reason why the coarse-graining modeling of diblock copolymers is, in general, more complicated than for homopolymers lies in the fact that a diblock copolymer is made of two different blocks, that for simplicity we label A and B. The two blocks may interact differently with the solvent and the scaling laws for A and B may therefore differ.

We need to consider several different length scales. First of all the two blocks can obey different scaling laws. Secondly, by splitting the chain into n subsegments, each segment is described by a radius of gyration that depends on n.

Another crucial characteristic of the system is the asymmetry ratio f between the two strands. The diblock copolymers considered have length L; they are made of a block A of length L_A and of a second block B of length L_B. Let R_{gA} be the radius of gyration of the A block and R_{gB} the radius of gyration of the B block. The scaling laws linking the radii of gyration to the number of monomers in the two branches are given by: $R_{gA} \sim L_A^{\nu_A}$, $R_{gB} \sim L_B^{\nu_B}$ with $\nu_A \neq \nu_B$.

We can then define the asymmetry ratio f in terms of the lengths L_A and L_B as:

$$f = \frac{L_A}{L_A + L_B} = \frac{L_A}{L}, \tag{1.12}$$

where we assumed that $b_A = b_B$.

We need to carefully develop a coarse-graining mechanism that preserves all the main properties, such as asymmetry ratio and the scaling laws. In order to determine the number of blobs needed to represent the polymer at each density, we make the assumption that the whole polymer behaves as if it were a homopolymer in good solvent. Furthermore, to compute the number of blobs needed at density $\rho = \alpha\rho^*$ (where ρ^* is the overlap density of the copolymer coils, and $\alpha > 1$), we assume that the polymer behaves as if it were a homopolymer in good solvent. In this way, we have, on the one hand, an overestimate of the number of blobs needed (since the radius of gyration of a homopolymer in good solvent is larger than that of a diblock copolymer), but on the other hand this assumption simplifies the description.

Following the approach to describe the homopolymers presented in Sect. 1.3, for a given density $\rho = \alpha\rho^*$, we have to use at least

$$n = n_A + n_B = \alpha^{1/(3\nu-1)} \tag{1.13}$$

blobs, where n_A and n_B are respectively the numbers of blobs that we use to represent the A and B parts of the chain and ν is the Flory exponent of a chain in a good solvent.

The next step is to determine the numbers of A and B blobs that is used to represent the chain. To this end, we must satisfy Eq. (1.12):

$$f = \frac{L_A}{L_A + L_B} = \frac{l_A n_A}{l_A n_A + l_B n_B}. \tag{1.14}$$

In the next section we present the solution adopted to define the model that can be used to describe the system.

1.5.1 The Equal Radii Solution of the Multi-Blob Model

To find the most suitable representation of the system, we start by explicitly rewriting Eq. (1.14) in terms of the number of blobs n_A, n_B, and n.

We wish to be able to simulate a certain degree of asymmetry f defined (in terms of the blobs) as

$$f = \frac{L_A}{L_A + L_B} = \frac{n_A \left(\frac{r_{gA}}{b_A}\right)^{1/\nu_A}}{n_A \left(\frac{r_{gA}}{b_A}\right)^{1/\nu_A} + n_B \left(\frac{r_{gB}}{b_B}\right)^{1/\nu_B}} \tag{1.15}$$

In Eq. (1.15), we have used $r_{gA} \simeq b_A l_A^{\nu_A} = b_A \left(\frac{L_A}{n_A}\right)^{\nu_A} \Rightarrow L_A = n_A \left(\frac{r_{gA}}{b_A}\right)^{1/\nu_A}$ and $r_{gB} \simeq b_B l_B^{\nu_B} = b_B \left(\frac{L_B}{n_B}\right)^{\nu_B} \Rightarrow L_B = n_B \left(\frac{r_{gB}}{b_B}\right)^{1/\nu_B}$, where r_{gA} is the radius of gyration of an A-blob; r_{gB} is that of a B-blob, and n_α is the number of blobs of type $\alpha = A, B$.

As we explore increasing densities, and consequently increase the number of blobs used to represent the polymer, we need to ensure that we do not change the asymmetry ratio.

Without any restriction, we may assume that the radii of gyration of the blobs used to represent the two blocks are equal. Therefore, if $r_{gA} = r_{gB}$, the asymmetry information in Eq. (1.15) is controlled by the number of blobs n_A and n_B needed to represent the A and the B blocks. By using the equal radii solution, the system is completely described by Eq. (1.15) with the extra condition $n_A + n_B = n$, where n has been computed according to Eq. (1.13).

1.5.2 Computing Effective Potentials for Diblock Copolymers

Moving one step forward into the coarse-graining strategy from the simple homopolymer case, we divide the system into a number $n = n_A + n_B$ of blobs. As explained in the previous section, all the blobs have the same radius of gyration; the total number of blobs is related to the density of polymers in solution and n_A and n_B determine the asymmetry ratio information.

To explain how to compute the potentials, we schematically divide the co-polymer into three regions:

- the first region is made by the $n_A - 1$ blobs that are not directly linked to a good solvent B blob. We call this region the "bad solvent zone."
- The second region is made by the bad solvent blob and the good solvent blob directly linked to it. We call this region the "diblock zone."
- The third region contains all the remaining $n_B - 1$ good solvent blobs that are not directly linked to a bad solvent blob. We refer to this region as the "good solvent zone " (see Fig. 1.9).

To compute the interactions between blobs belonging to two distinct chains, we split the chains into three types of dimer, as sketched in Fig. 1.9 and, as explained in Fig. 1.10, extracting the potentials between two different dimers is a four-body problem.

It hence appears that, in order to fully describe the interactions in the system, we have to compute, for example, the potentials between a dimer belonging to the "bad solvent zone" interacting with another "bad solvent zone" dimer, then with a

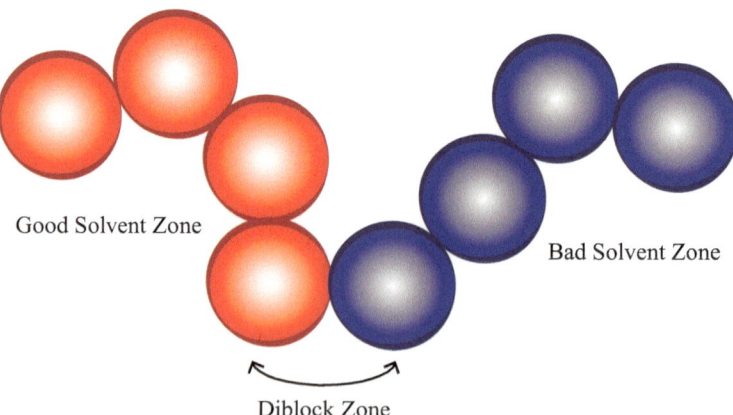

Fig. 1.9 A schematic representation of the three zones we have to divide the polymer into, in order to compute the effective potentials between the blobs that we use in the coarse-grained representation. The blue spheres represent the bad solvent blobs, while the red ones represent the good solvent ones

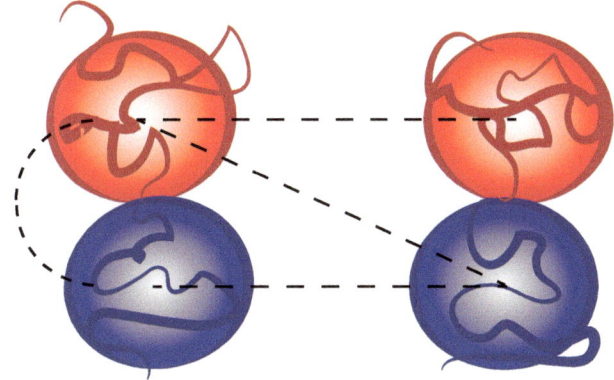

Fig. 1.10 We split the polymer represented in Fig. 1.9 into three dimer like parts. We run separate simulations of these sub-parts of the polymer in order to extract the effective potentials needed for a coarse-grained representation of the polymers. The interactions between two dimers are sketched in this plot. Each blob, when interacting with a blob belonging to the other chain, is tethered to a second blob which is also interacting with the second chain

Table 1.1 The interactions between blobs belonging to different "zones"

	$A^{\text{Bad solvent}}$	A^{Diblock}	B^{Diblock}	$B^{\text{Good solvent}}$
$A^{\text{Bad solvent}}$	$v_{AA}^{\text{B-B}}$	$v_{AA}^{\text{B-D}}$	$v_{AB}^{\text{B-D}}$	$v_{AB}^{\text{B-G}}$
A^{Diblock}	$v_{AA}^{\text{D-B}}$	v_{AA}^{D}	v_{AB}^{D}	$v_{AB}^{\text{D-G}}$
B^{Diblock}	$v_{BA}^{\text{D-B}}$	v_{BA}^{D}	v_{BB}^{D}	$v_{BB}^{\text{D-G}}$
$B^{\text{Good solvent}}$	$v_{BA}^{\text{G-B}}$	$v_{BA}^{\text{G-D}}$	$v_{BB}^{\text{G-D}}$	$v_{BB}^{\text{G-G}}$

In the table, D stands for diblock, B for bad solvent, and G for good solvent

"diblock zone" dimer and with a "good-solvent zone" dimer. A similar procedure can be described for all the dimers we split the original chain into.

We list in Table 1.1 all the possible intermolecular interactions that each blob, belonging to one of the three zones, can experience while interacting with a blob belonging to the same or a different polymer. The table is symmetric. A priori the total number of interactions between two polymers involves ten independent intermolecular potentials and three distinct intramolecular tethering potentials $\varphi_{AA}^{\text{Bad solvent}}$, $\varphi_{AB}^{\text{Diblock}}$, and $\varphi_{BB}^{\text{Good solvent}}$.

We proceed as we did both for the dumbbell representation of diblock copolymers and for the multi-blob representation of homopolymers, to compute the interaction potentials in the low density limit.

Full monomer conformations of two isolated polymer dimers are generated, and the CMs of the two α or β blobs are moved towards each other. Histograms of the allowed configurations are collected as a function of the CM$_\alpha$–CM$_\beta$ distances, with $\alpha, \beta \in [A, B]$. By defining $h_{\alpha\beta}(r) = g_{\alpha\beta}(r) - 1$ the intermolecular, and $s_{\alpha\beta}(r)$ the intramolecular pair distribution function, we can write down the equations that link the pair distribution functions to the effective potentials [40, 87].

Let us define the following quantities:

$$h_{\alpha\beta}(r) = \lim_{\rho \to 0} h_{\alpha\beta}(r)$$

$$s_{\alpha\beta}(r) = \exp\left[-\frac{\varphi_{\alpha\beta}(r)}{k_B T}\right] \tag{1.16}$$

$$f_{\alpha\beta}(r) = \exp\left[-\frac{v_{\alpha\beta}(r)}{k_B T}\right] - 1.$$

There are six possible combinations of dimers made of A ($\circ$) and B ($\bullet$) monomers:

1. $\left.\begin{array}{c} \circ - \circ \\ \circ - \circ \end{array}\right\} \Rightarrow h_{AA}(r)$

This combination involves the simulation of two monomer-resolved dumb-bells made of A monomers in bad solvent. Since the bad solvent monomers do not interact, the effective potential between two bad solvent dimers is zero. The only potential we extract by means of a simulation of a single dimer is the tethering potential $\varphi_{AA}(r)$ between the two blobs in bad solvent, which is known exactly (Gaussian chain).

2. $\left.\begin{array}{c} \circ - \circ \\ \circ - \bullet \end{array}\right\} \Rightarrow h_{AA}(r), h_{AB}(r)$

From the inversion of the pair distribution functions between an A–B dimer and an A–A dimer, we can obtain two effective potentials, namely $v_{AA}(r)$, acting between the centers of mass of the blobs in bad solvent and $v_{AB}(r)$, acting between the CMs of non-tethered A and B blobs.

The exact expressions for $h_{AA}(r)$ and $h_{AB}(r)$ in terms of the $s_{\alpha\beta}$ and $f_{\alpha\beta}$ are:

$$h_{AA}(\mathbf{r}) = f_{AA}(\mathbf{r}) + [1 + f_{AA}(\mathbf{r})]\left\{\int [d\mathbf{x} f_{AA}(\mathbf{x})s_{AA}(\mathbf{x} - \mathbf{r}) + s_{AB}(\mathbf{x})f_{BA}(\mathbf{x} - \mathbf{r})]\right.$$

$$+ \int d\mathbf{x} \int d\mathbf{y}\, s_{AA}(\mathbf{x})s_{BA}(\mathbf{y} - \mathbf{r})\, [f_{AB}(\mathbf{x} - \mathbf{y}) + f_{BA}(\mathbf{y})f_{AB}(\mathbf{x} - \mathbf{y})$$

$$\left. + f_{AA}(\mathbf{x} - \mathbf{r})f_{AB}(\mathbf{x} - \mathbf{y}) + f_{BA}(\mathbf{y})f_{AA}(\mathbf{x} - \mathbf{r}) + f_{BA}(\mathbf{y})f_{AA}(\mathbf{x} - \mathbf{r})f_{AB}(\mathbf{x} - \mathbf{y})]\right\}$$

$$h_{AB}(\mathbf{r}) = f_{AB}(\mathbf{r}) + [1 + f_{AB}(\mathbf{r})]\left\{\int [d\mathbf{x} f_{AA}(\mathbf{x})s_{AB}(\mathbf{x} - \mathbf{r}) + s_{AA}(\mathbf{x})f_{AB}(\mathbf{x} - \mathbf{r})]\right.$$

$$+ \int d\mathbf{x} \int d\mathbf{y}\, s_{AA}(\mathbf{x})s_{AB}(\mathbf{y} - \mathbf{r})\, [f_{AA}(\mathbf{x} - \mathbf{y}) + f_{AA}(\mathbf{y})f_{AA}(\mathbf{x} - \mathbf{y})$$

$$\left. + f_{AB}(\mathbf{x} - \mathbf{r})f_{AA}(\mathbf{x} - \mathbf{y}) + f_{AA}(\mathbf{y})f_{AB}(\mathbf{x} - \mathbf{r}) + f_{AA}(\mathbf{y})f_{AB}(\mathbf{x} - \mathbf{r})f_{AA}(\mathbf{x} - \mathbf{y})]\right\}$$

$$\tag{1.17}$$

3. $\left.\begin{array}{l} \circ - \circ \\ \bullet - \bullet \end{array}\right\} \Rightarrow h_{AB}(r)$

These simulations measure the pair distribution functions used to obtain the interactions between dimers belonging to the bad solvent zone and dimers belonging to the good solvent zone. We show in the following that the resulting effective potential $v_{AB}(r)$ is almost identical to the effective potential $v_{AB}(r)$ computed in cases 2, 4, and 5.

$$h_{AB}(\mathbf{r}) = f_{AB}(\mathbf{r}) + [1 + f_{AB}(\mathbf{r})] \left\{ \int [d\mathbf{x} f_{AB}(\mathbf{x}) s_{BB}(\mathbf{x} - \mathbf{r}) + s_{AA}(\mathbf{x}) f_{AB}(\mathbf{x} - \mathbf{r})] \right.$$

$$+ \int d\mathbf{x} \int d\mathbf{y}\; s_{AA}(\mathbf{x}) s_{BB}(\mathbf{y} - \mathbf{r}) [f_{AB}(\mathbf{x} - \mathbf{y}) + f_{AB}(\mathbf{y}) f_{AB}(\mathbf{x} - \mathbf{y})$$

$$\left. + f_{AB}(\mathbf{x} - \mathbf{r}) f_{AB}(\mathbf{x} - \mathbf{y}) + f_{AB}(\mathbf{y}) f_{AB}(\mathbf{x} - \mathbf{r}) + f_{AB}(\mathbf{y}) f_{AB}(\mathbf{x} - \mathbf{r}) f_{AB}(\mathbf{x} - \mathbf{y})] \right\}$$

$$(1.18)$$

4. $\left.\begin{array}{l} \circ - \bullet \\ \circ - \bullet \end{array}\right\} \Rightarrow h_{AA}(r), h_{AB}(r), h_{BB}(r)$

In the case of two A–B dimers, the equations read:

$$h_{AA}(\mathbf{r}) = f_{AA}(\mathbf{r}) + [1 + f_{AA}(\mathbf{r})] \left\{ \int [d\mathbf{x} f_{AB}(\mathbf{x}) s_{BA}(\mathbf{x} - \mathbf{r}) + s_{AB}(\mathbf{x}) f_{BA}(\mathbf{x} - \mathbf{r})] \right.$$

$$+ \int d\mathbf{x} \int d\mathbf{y}\; s_{AB}(\mathbf{x}) s_{BA}(\mathbf{y} - \mathbf{r}) [f_{BB}(\mathbf{x} - \mathbf{y}) + f_{AB}(\mathbf{y}) f_{BB}(\mathbf{x} - \mathbf{y})$$

$$\left. + f_{BA}(\mathbf{x} - \mathbf{r}) f_{BB}(\mathbf{x} - \mathbf{y}) + f_{AB}(\mathbf{y}) f_{BA}(\mathbf{x} - \mathbf{r}) + f_{AB}(\mathbf{y}) f_{BA}(\mathbf{x} - \mathbf{r}) f_{BB}(\mathbf{x} - \mathbf{y})] \right\}$$

$$h_{AB}(\mathbf{r}) = f_{AB}(\mathbf{r}) + [1 + f_{AB}(\mathbf{r})] \left\{ \int [d\mathbf{x} f_{AA}(\mathbf{x}) s_{AB}(\mathbf{x} - \mathbf{r}) + s_{AB}(\mathbf{x}) f_{BB}(\mathbf{x} - \mathbf{r})] \right.$$

$$+ \int d\mathbf{x} \int d\mathbf{y}\; s_{AB}(\mathbf{x}) s_{AB}(\mathbf{y} - \mathbf{r}) [f_{BA}(\mathbf{x} - \mathbf{y}) + f_{AA}(\mathbf{y}) f_{BA}(\mathbf{x} - \mathbf{y})$$

$$\left. + f_{BB}(\mathbf{x} - \mathbf{r}) f_{BA}(\mathbf{x} - \mathbf{y}) + f_{AA}(\mathbf{y}) f_{BB}(\mathbf{x} - \mathbf{r}) + f_{AA}(\mathbf{y}) f_{BB}(\mathbf{x} - \mathbf{r}) f_{BA}(\mathbf{x} - \mathbf{y})] \right\}$$

$$h_{BB}(\mathbf{r}) = f_{BB}(\mathbf{r}) + [1 + f_{BB}(\mathbf{r})] \left\{ \int [d\mathbf{x} f_{BA}(\mathbf{x}) s_{AB}(\mathbf{x} - \mathbf{r}) + s_{BA}(\mathbf{x}) f_{AB}(\mathbf{x} - \mathbf{r})] \right.$$

$$+ \int d\mathbf{x} \int d\mathbf{y}\; s_{BA}(\mathbf{x}) s_{AB}(\mathbf{y} - \mathbf{r}) [f_{Aa}(\mathbf{x} - \mathbf{y}) + f_{BA}(\mathbf{y}) f_{AA}(\mathbf{x} - \mathbf{y})$$

$$\left. + f_{AB}(\mathbf{x} - \mathbf{r}) f_{AA}(\mathbf{x} - \mathbf{y}) + f_{BA}(\mathbf{y}) f_{AB}(\mathbf{x} - \mathbf{r}) + f_{BA}(\mathbf{y}) f_{AB}(\mathbf{x} - \mathbf{r}) f_{AA}(\mathbf{x} - \mathbf{y})] \right\}$$

$$(1.19)$$

5. $\genfrac{}{}{0pt}{}{\circ - \bullet}{\bullet - \bullet}\Bigg\} \Rightarrow h_{AB}(r), h_{BB}(r)$

The equations obtained for the potentials in this case are analogous to what was obtained in case 2 of this list, by simply exchanging A with B:

$$h_{AB}(\mathbf{r}) = f_{BA}(\mathbf{r}) + [1 + f_{BA}(\mathbf{r})] \left\{ \int [d\mathbf{x} f_{BB}(\mathbf{x}) s_{BA}(\mathbf{x} - \mathbf{r}) + s_{BB}(\mathbf{x}) f_{BA}(\mathbf{x} - \mathbf{r})] \right.$$

$$+ \int d\mathbf{x} \int d\mathbf{y}\, s_{BB}(\mathbf{x}) s_{BA}(\mathbf{y} - \mathbf{r}) [f_{BB}(\mathbf{x} - \mathbf{y}) + f_{BB}(\mathbf{y}) f_{BB}(\mathbf{x} - \mathbf{y})$$

$$\left. + f_{BA}(\mathbf{x} - \mathbf{r}) f_{BB}(\mathbf{x} - \mathbf{y}) + f_{BB}(\mathbf{y}) f_{BA}(\mathbf{x} - \mathbf{r}) + f_{BB}(\mathbf{y}) f_{BA}(\mathbf{x} - \mathbf{r}) f_{BB}(\mathbf{x} - \mathbf{y})] \right\}$$

$$h_{BB}(\mathbf{r}) = f_{BB}(\mathbf{r}) + [1 + f_{BB}(\mathbf{r})] \left\{ \int [d\mathbf{x} f_{BB}(\mathbf{x}) s_{BB}(\mathbf{x} - \mathbf{r}) + s_{BA}(\mathbf{x}) f_{AB}(\mathbf{x} - \mathbf{r})] \right.$$

$$+ \int d\mathbf{x} \int d\mathbf{y}\, s_{BB}(\mathbf{x}) s_{AB}(\mathbf{y} - \mathbf{r}) [f_{BA}(\mathbf{x} - \mathbf{y}) + f_{BA}(\mathbf{y}) f_{BA}(\mathbf{x} - \mathbf{y})$$

$$\left. + f_{BB}(\mathbf{x} - \mathbf{r}) f_{BA}(\mathbf{x} - \mathbf{y}) + f_{AB}(\mathbf{y}) f_{BB}(\mathbf{x} - \mathbf{r}) + f_{AB}(\mathbf{y}) f_{BB}(\mathbf{x} - \mathbf{r}) f_{BA}(\mathbf{x} - \mathbf{y})] \right\}$$

$$\tag{1.20}$$

6. $\genfrac{}{}{0pt}{}{\bullet - \bullet}{\bullet - \bullet}\Bigg\} \Rightarrow h_{BB}(r)$

The last inversion listed gives the effective potential between two non-tethered good solvent blobs. The equation that we invert to obtain the effective potential reads:

$$h_{BB}(\mathbf{r}) = f_{BB}(\mathbf{r}) + [1 + f_{BB}(\mathbf{r})] \left\{ \int [d\mathbf{x} f_{BB}(\mathbf{x}) s_{BB}(\mathbf{x} - \mathbf{r}) + s_{BB}(\mathbf{x}) f_{BB}(\mathbf{x} - \mathbf{r})] \right.$$

$$+ \int d\mathbf{x} \int d\mathbf{y}\, s_{BB}(\mathbf{x}) s_{BB}(\mathbf{y} - \mathbf{r}) [f_{BB}(\mathbf{x} - \mathbf{y}) + f_{BB}(\mathbf{y}) f_{BB}(\mathbf{x} - \mathbf{y})$$

$$\left. + f_{BB}(\mathbf{x} - \mathbf{r}) f_{BB}(\mathbf{x} - \mathbf{y}) + f_{BB}(\mathbf{y}) f_{BB}(\mathbf{x} - \mathbf{r}) + f_{BB}(\mathbf{y}) f_{BA}(\mathbf{x} - \mathbf{r}) f_{BB}(\mathbf{x} - \mathbf{y})] \right\}$$

$$\tag{1.21}$$

We now have to calculate the length of the different parts that constitute each dimer, in order to compute the effective potentials. Fixing $r_{gA} = r_{gB}$ implies that $l_A = (b_B/b_A)^{1/\nu_A} l_B^{\nu_B/\nu_A}$; for a given length l_A, l_B therefore follows, e.g. if $l_B = 300$, $l_A = (b_B/b_A)^{1/\nu_A} \cdot 300^{\nu_B/\nu_A} = 950$.

Once the two lengths l_A and l_B are known, we perform the different simulations to obtain all the desired potentials.

Due to the equal radii approach, the potentials that we are computing are independent of the asymmetry ratio f. Thereby we obtain a universal set of potentials that we use to study different values of f at different densities $\rho = \alpha \rho^*$ with $\alpha > 1$ in the semi-dilute regime, by simply changing the numbers n_A and n_B.

1.5.3 Summary of the Model

By imposing the condition $r_{gA} = r_{gB}$, we can compute once and for all the effective potentials from a microscopic point of view and use them to explore a wide range of asymmetry ratios f. The potentials between blobs of the same kind appear to be weakly influenced by the identity of the blob they are linked to. This principle holds for all the potentials except for the v_{AA} case. The strongest repulsion between two bad solvent blobs is the one coming from the interaction of two "diblock zone" dimers. Although this repulsion is almost 20 times weaker than the v_{BB} or than the v_{AB} repulsion, we still include it explicitly in the description of our model, while we consider negligible all the other v_{AA} potentials. Hence, we have reduced the total number of distinct (intermolecular and intramolecular) potentials to just seven pair interactions, one of which is identically zero.

The equal radii solution of the multi-blob approach allows us to compute the potentials at very low density. By then fixing the total number of blobs used to represent density $\tilde{\rho} = \rho/\rho^*$ by the relation $n = n_A + n_B = (\rho/\rho^*)^{\frac{1}{3\nu-1}}$, it is hence possible to represent diblock copolymers of distinct asymmetries by simply changing the ratio n_A/n_B according to Eq. (1.14).

This approach avoids any kind of rescaling of the potentials when increasing the number of blobs since, by increasing $n = n_A + n_B$ and using the information carried in Eq. (1.15), the rescaling of the interaction is imposed changing the number of blobs. In other words, when we increase the total number of blobs n, for a fixed f we compute the number of blobs n_A and n_B while keeping fixed $r_{gA} = r_{gB}$. The radii r_{gA} and r_{gB} are of course changing as n increases, but their ratio remains $r_{gA}/r_{gB} = 1$.

The model introduced in this section is a powerful tool that, for example, allowed to explore, with a single set of potentials, the $(f-\rho)$ phase diagram and to quantitatively compare the results to experimental data [37].

1.6 Self-Assembly and Hierarchical Self-Assembly: Diblock Copolymers and Telechelic Star Polymers

Extending the multiscale coarse graining strategy to first deal with topological constraints and to then include different sets of interactions allows to analyze the self-assembly properties of macromolecules characterized by both complex sets of interactions or complex architecture.

It thus becomes feasible to follow the lead of experimental work [84, 85, 88, 89] and of theoretical predictions [90, 91] to first explore the assembling properties of diblock copolymers, and then to design novel macromolecules, e.g. diblock copolymer stars, to exploit the entropic/enthalpic competition arising within each diblock copolymer arm.

The accuracy of the exploration of a polymeric phase diagram, deep into the semi-dilute regime, is strongly related to the accuracy with which effective potential used describes single molecule properties: the development of a back-trackable coarse-graining technique that takes into account the roles of many-body interactions has been proven to qualitatively and almost quantitatively reproduce the correct phase diagram measured in experiments [37].

As mentioned, the radius of gyration R_g^b of a multi-blob chain differs from the radius of gyration R_g of the underlying microscopic model, therefore implying that:

$$\rho_b^* = \frac{3}{4\pi R_g^{b\,3}} \neq \rho^* = \frac{3}{4\pi R_g^3}. \tag{1.22}$$

To map the results obtained with the coarse-grained simulations onto the microscopic experimental results in [84, 85, 88, 89], a correct rescaling of the radius of gyration of the multi-blob chains is needed, that projects onto a rescaling for the densities.

$$\frac{\rho}{\rho^*} = \frac{\rho_b}{\rho_b^*} \Rightarrow \rho = \rho_b \frac{\rho^*}{\rho_b^*}$$

$$\rho = \rho_b \left(\frac{R_g^b}{R_g}\right)^3 \sim \frac{\rho_b}{2}. \tag{1.23}$$

1.7 Quantitative Exploration of the Semi-dilute Regime with a Multi-Blob Approach

A quantitative and simple to handle multi-scale coarse-graining strategy allows, by quantitatively exploring the semi-dilute regime, to simply compute the free energies relative to different equilibrium phases and to define its phase transitions.

The free energy F is linked to the chemical potential and the pressure by the relation:

$$F = N\mu - PV, \tag{1.24}$$

where μ is the chemical potential of the copolymers in solution and P the osmotic pressure.

In order to compute the excess free energy for the system, Monte Carlo *NVT* simulations can be–for example–performed to calculate the excess pressure of the system and the excess chemical potential. To describe the derivation in full, in the following section we briefly analyze the statistical mechanics of a fluid of diblock copolymers.

The total partition function of the low density system can be easily written as the total partition function of a gas of N independent polymers: $Q_N^0 = q^N/N!$, where

$$
\begin{aligned}
q &= \int \prod_{k=1}^{n_A} \prod_{l=1}^{n_B} d\mathbf{p}_k^A \, d\mathbf{p}_l^B \exp\left\{-\beta\left[\frac{(\mathbf{p}_k^A)^2}{2m_A} + \frac{(\mathbf{p}_l^B)^2}{2m_B}\right]\right\} \\
&\quad \cdot \int d^{n_A}\mathbf{r}^A \int d^{n_B}\mathbf{r}^B \, \exp\left[-\beta U\left(\{\mathbf{r}^A\}^{(n_A)}, \{\mathbf{r}^B\}^{(n_B)}\right)\right] \\
&= \int \prod_{k=1}^{n_A} \prod_{l=1}^{n_B} d\mathbf{p}_k^A \, d\mathbf{p}_l^B \exp\left[-\beta\frac{(\mathbf{p}_k^A)^2}{2m_A} + \frac{(\mathbf{p}_l^B)^2}{2m_B}\right] \\
&\quad \cdot \int d\mathbf{r}_1^A \int d^{(n_A-1)}\mathbf{s}^A \, d^{(n_B)}\mathbf{s}^B \, \exp\left[-\beta U^{\mathtt{Intra}}\left(\{\mathbf{s}^A\}^{(n_A-1)}, \{\mathbf{s}^B\}^{(n_B)}\right)\right] \\
&= \frac{V}{\Lambda_A^{3n_A}\Lambda_B^{3n_B}} I(\beta)
\end{aligned}
$$

$$(1.25)$$

where:

$$
\begin{aligned}
U^{\mathtt{Intra}} &= \sum_{i=1}^{n_A-1} \varphi_{AA}(|\mathbf{r}_{i+1}^A - \mathbf{r}_i^A|) + \varphi_{AB}(|\mathbf{r}_1^B - \mathbf{r}_{n_A}^A|) + \sum_{i=1}^{n_B-1} \varphi_{BB}(|\mathbf{r}_{i+1}^B - \mathbf{r}_i^B|) \\
&\quad + \sum_{i=1}^{n_A}\sum_{j>i+1}^{n_A} v_{AA}(|\mathbf{r}_j^A - \mathbf{r}_i^A|) + \sum_{i=1}^{n_B}\sum_{j>i+1}^{n_B} v_{BB}(|\mathbf{r}_j^B - \mathbf{r}_i^B|) \\
&\quad + \sum_{i=1}^{n_A}\sum_{j=1}^{n_B} v_{AB}(|\mathbf{r}_j^B - \mathbf{r}_i^A|) - v_{AB}(|\mathbf{r}_1^B - \mathbf{r}_{n_a}^A|)
\end{aligned}
$$

$$
\Lambda_{(A,B)} = \sqrt{\frac{h^2}{2\pi m_{(A,B)} k_B T}}
$$

$$
\mathbf{s}_i^A = \mathbf{r}_i^A - \mathbf{r}_1^A \text{ for } i \in [2, n_A]
$$

$$
\mathbf{s}_i^B = \mathbf{r}_i^B - \mathbf{r}_1^A \text{ for } i \in [1, n_B]
$$

and

$$
I(\beta) = \int d\mathbf{r}_1^A \int d^{(n_A-1)}\mathbf{s}^A \, d^{(n_B)}\mathbf{s}^B \, \exp\left[-\beta U^{\mathtt{Intra}}\left(\{\mathbf{s}^A\}^{(n_A-1)}, \{\mathbf{s}^B\}^{(n_B)}\right)\right]
$$

$$(1.26)$$

$U^{\mathtt{Intra}}$ is the intramolecular potential of a single polymer, n_A and n_B are the total number of A and B blobs, $\mathbf{r}_i^{A,B}$ are the coordinates of the centers of mass of the i-th blobs, $\mathbf{p}^A$ and $\mathbf{p}^B$ the momenta, respectively, of the monomers A and B and $I(\beta)$ is a quantity that only depends on the temperature and on the solvent quality. Let us introduce $\Lambda = \Lambda_A^{n_A} \Lambda_B^{n_B}$, we can therefore write the total partition function at low density for a gas of non-interacting diblock copolymers and obtain:

$$Q_N^0 = \frac{V^N I(\beta)^N}{N! \Lambda^{3N}} \tag{1.27}$$

From (1.27) we can extract both the ideal free energy, the ideal chemical potential and the ideal pressure from the relations

$$F^0 = -k_B T \ln Q^0 = N k_B T \left(\ln(N/V) - 1 - \ln\left(\frac{I(\beta)}{\Lambda^3}\right) \right)$$

$$P^0 = \left(-\frac{\partial F^0}{\partial V} \right)_T = \frac{N k_B T}{V} \tag{1.28}$$

$$\mu^0 = \frac{\partial F^0}{\partial N} = k_B T \left[\ln(N/V) - \ln\left(\frac{I(\beta)}{\Lambda^3}\right) \right]$$

Once obtained the partition function for low density, non-interacting diblock copolymers, we can write the total partition function of the system at finite density. The polymers interact with a total potential energy U_N that includes both the interactions between monomers (or blobs) belonging to two different polymers and the tethering interactions.

$$U_N = \sum_{k=1}^{N} U_k^{\texttt{Intra}} + \sum_{k=1}^{N} \sum_{l=1}^{N} U_{lk}^{\texttt{Inter}} \tag{1.29}$$

where $U_k^{\texttt{Intra}}$ was defined in Eq. (1.26) for a single chain and $U_{lk}^{\texttt{Inter}}$ is defined as

$$U_{lk}^{\texttt{Inter}} = \sum_{i=1}^{n_A} \sum_{j}^{n_A} v_{AA}(|\mathbf{r}_j^{A_k} - \mathbf{r}_i^{A_l}|) + \sum_{i=1}^{n_B} \sum_{j}^{n_B} v_{BB}(|\mathbf{r}_j^{B_k} - \mathbf{r}_i^{B_l}|) + \sum_{i=1}^{n_A} \sum_{j=1}^{n_B} v_{AB}(|\mathbf{r}_j^{B_k} - \mathbf{r}_i^{A_l}|)$$

$$\tag{1.30}$$

We can write the total partition function as:

$$Q_N = \frac{1}{N! \Lambda^{3N}} \int d^{N n_A} \mathbf{r}_A d^{N n_B} \mathbf{r}_B \exp\left(-\beta U_N\right) = Q_N^0 Z_N \tag{1.31}$$

where

$$Z_N = \frac{1}{V^N I(\beta)^N} \int d^{N n_A} \mathbf{r}_A d^{N n_B} \mathbf{r}_B \exp\left(-\beta U_N\right) \tag{1.32}$$

N being the total number of polymers in solution and V the volume.

If we now assume to be in a cubic box of box length L, the volume V is $V = L^3$. We can hence define a dimensionless coordinate system such that:

$$\mathbf{r}_i^A = L\mathbf{u}_i^A \tag{1.33}$$

$$\mathbf{r}_i^B = L\mathbf{u}_i^B. \tag{1.34}$$

If we express the partition function Z_N as a function of the scaled coordinate system, we get:

$$
\begin{aligned}
Z_N &= \frac{L^{3n_A N} L^{3n_B N}}{V^N I^N(\beta)} \int d^{Nn_A} \mathbf{u}_A d^{Nn_B} \mathbf{u}_B \exp\left(-\beta U_N\right) \\
&= \frac{V^{N(n-1)}}{I^N(\beta)} \Gamma
\end{aligned}
\tag{1.35}
$$

where $n = n_A + n_B$ and

$$\Gamma = \int d^{Nn_A} \mathbf{u}_A d^{Nn_B} \mathbf{u}_B \exp\left(-\beta U_N\right). \tag{1.36}$$

The total free energy of the system at finite density is equal to:

$$F = -k_B T \ln(Q_N) = F^0 + F^{\text{ex}}, \quad \text{where} \quad F^{\text{ex}} = -k_B T \ln(Z_N) \tag{1.37}$$

Having defined the excess free energy F^{ex}, we can compute the excess pressure P^{ex}:

$$\beta P^{\text{ex}} = \left(-\frac{\partial \beta F^{\text{ex}}}{\partial V}\right)_T = \frac{1}{Z_N}\left(\frac{\partial Z_N}{\partial V}\right)_T. \tag{1.38}$$

We can compute the derivative of the partition function Z_N in Eq. (1.35) with respect to the volume V:

$$
\begin{aligned}
\frac{\partial Z_N}{\partial V} &= \frac{N(n-1)V^{N(n-1)-1}}{I^N(\beta)}\Gamma + \left(\frac{V^{N(n-1)}}{I^N(\beta)}\right)\frac{1}{3L^2}\frac{\partial \Gamma}{\partial L} \\
&= \frac{N(n-1)V^{N(n-1)-1}}{I^N(\beta)}\Gamma - \left(\frac{\beta}{I^N(\beta)3V}\right)\int d^{n_A N}\mathbf{r}_A d^{n_B N}\mathbf{r}_B \exp\left(-\beta U_N\right)\nabla U_N \cdot \vec{r} \\
&= \frac{N(n-1)}{V}Z_N + \frac{\beta}{3VI^N(\beta)}\int d^{n_A N}\mathbf{r}_A d^{n_B N}\mathbf{r}_B \exp\left(-\beta U_N\right)\sum_{i<j}\vec{f}(\vec{r}_{ij})\cdot\vec{r}_{ij},
\end{aligned}
\tag{1.39}
$$

where $\vec{f}(\vec{r}_{ij})$ is the force between the particles i and j at a distance $\vec{r}_{ij}$. Gathering the results, the excess pressure is:

$$\beta P^{\text{ex}} = \frac{N(n-1)}{V} + \frac{\beta}{3V}\left\langle \sum_{i<j} \vec{f}(\vec{r}_{ij}) \cdot \vec{r}_{ij} \right\rangle, \tag{1.40}$$

and the total pressure of the system is

$$\beta P = \beta\left(P^0 + P^{\text{ex}}\right) = \frac{Nn}{V} + \frac{\beta}{3V}\left\langle \sum_{i<j} \vec{f}(\vec{r}_{ij}) \cdot \vec{r}_{ij} \right\rangle. \tag{1.41}$$

The chemical potential can be extracted as well from the total free energy of the system and, as for the pressure, we split it into a contribution coming from a low density system μ^0 and an excess part μ^{ex}.

$$\beta\mu = \frac{\partial \ln\left(Q^0 Z_N\right)}{\partial N} = \frac{\partial}{\partial N}\left[\ln\left(Q^0\right) + \ln\left(Z_N\right)\right] = \beta\mu^0 + \frac{\partial \ln\left(Z_N\right)}{\partial N} = \beta\mu^0 + \ln\left(\frac{Z_{N+1}}{Z_N}\right)$$

$$= \ln(N/V) + \ln\left(\frac{\Lambda^3}{I(\beta)}\right) - \ln\left(\frac{\int d^{Nn_A}\mathbf{r}^A d^{Nn_B}\mathbf{r}^B e^{(-\beta U_{N+1})}}{VI(\beta)\int d^{Nn_A}\mathbf{r}^A d^{Nn_B}\mathbf{r}^B e^{(-\beta U_N)}}\right)$$

$$= \ln(N/V) + \ln\left(\frac{\Lambda^3}{I(\beta)}\right) - \ln\left(\frac{1}{VI(\beta)}\frac{J_{N+1}}{J_N}\right)$$

$$\tag{1.42}$$

We can now write the potential terms in the two integrals J_{N+1} and J_N by highlighting the contribution to both the intramolecular and intermolecular potentials, due to the $(N+1)$-th polymer. By calling W the contribution to the total potential due to the $(N+1)$ polymer and separating the total potential energy into $U_N = U_N^{\text{Intra}} + U_N^{\text{Inter}}$, we can write:

$$\begin{cases} U_{N+1}^{\text{Intra}} = U_N^{\text{Intra}} + W^{\text{Intra}}(1) = U_N^{\text{Intra}} + U_1^{\text{Intra}} \\ U_{N+1}^{\text{Inter}} = U_N^{\text{Inter}} + W^{\text{Inter}}(N,1) \end{cases} \tag{1.43}$$

where the additional inserted particle is polymer 1.

The ratio J_{N+1}/J_N is hence:

$$\frac{J_{N+1}}{J_N} = \int d^{n_A}\mathbf{r}_1^A d^{n_B}\mathbf{r}_1^B e^{(-\beta U_1^{\text{Intra}})}\left\langle \exp\left(-\beta W^{\text{Inter}}(N,1)\right)\right\rangle_N \tag{1.44}$$

where $\langle \bullet \rangle_N$ is the canonical average of $\bullet$ over the N polymer ensemble weighted by the Boltzmann factor $e^{-\beta\left(U_N^{\text{Intra}}+U_N^{\text{Inter}}\right)}$. By applying translational invariance of the integral, choosing $\mathbf{r}_1^A$ (namely the coordinates of the first blob A of the polymer number 1) as the origin, we can write:

$$\frac{J_{N+1}}{J_N} = \int d\mathbf{r}_1^A \int d^{n_A-1}\mathbf{r}_1^A d^{n_B}\mathbf{r}_1^B e^{(-\beta U_1^{\text{Intra}})} \left\langle \exp\left(-\beta W^{\text{Inter}}(N,1)\right)\right\rangle_N$$
$$= V \int d^{n_A-1}\mathbf{r}_1^A d^{n_B}\mathbf{r}_1^B e^{(-\beta U_1^{\text{Intra}})} \left\langle \exp\left(-\beta W^{\text{Inter}}(N,1)\right)\right\rangle_N \tag{1.45}$$

The total chemical potential hence reads:

$$\beta\mu = \ln(N/V) + \ln\left(\Lambda^3\right) - \ln\left(\int d^{n_A-1}\mathbf{r}_1^A d^{n_B}\mathbf{r}_1^B e^{-\left(\beta U_1^{\text{Intra}}\right)} \left\langle e^{\left(-\beta W^{\text{Inter}}(N,1)\right)}\right\rangle_N\right) \tag{1.46}$$

The excess chemical potential of this equation can be—for example—computed in simulations by means of the Widom insertion method [92] by inserting in an NVT simulation a diblock copolymer, equilibrated in a gas reservoir of diblock copolymers, as a "ghost particle." In Eq. (1.46) $W^{\text{Inter}}(N,1)$ is the interaction of the test chain with the rest of the system. At frequent intervals during the simulation, we hence generate a chain according to the intramolecular Boltzmann weight and we randomly insert it in the simulation box. It is then possible to compute the intermolecular interaction between the "ghost" chain and the rest of the system. The excess chemical potential is computed by averaging $W^{\text{Inter}}(N,1)$ over all the configurations of the N polymers in the box.

We conclude this section by writing explicitly the total free energy of the system:

$$\beta F = \beta N\mu - \beta PVN(\ln(\rho) - n) - N\ln\int d^{Nn_A}\mathbf{r}_1^A d^{Nn_B}\mathbf{r}_1^B \left\langle e^{(-\beta\Delta U)}\right\rangle_N$$
$$+ \frac{\beta}{3}\left\langle\sum_{i<j}\vec{f}(\vec{r}_{ij})\cdot\vec{r}_{ij}\right\rangle + \ln(\Lambda^{3N}). \tag{1.47}$$

1.8 Playing with the Macromolecular Architecture: Chains and Stars

The development of a whole theoretical framework that allows for a multiscale representation of polymeric systems with the ever-present possibility of backtracking the results to full-monomer simulations and/or to an experimental realization of the system allows—as was already shown in the case of ring polymers—to deeply explore the semi-dilute regime in polymer solutions with the possibility of back tracking the results onto experiments and microscopic realization.

Diblock copolymers are known to self-assemble into ordered or disordered supra-molecular aggregates or microphases, including micelles, lamellae, cylinders or bicontinuous structures, both in the melt [76,93] and in selective solvents [75,84].

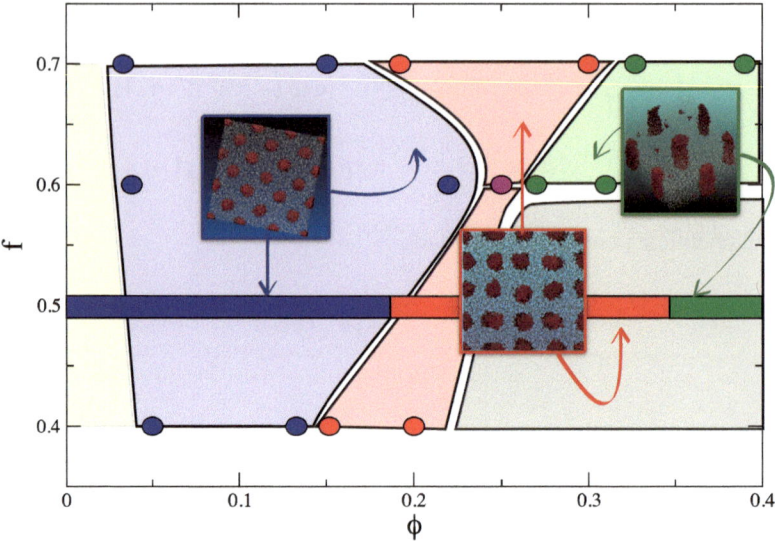

Fig. 1.11 Comparison between a computational phase diagram obtained for diblock copolymer solution in selective solvent by means of a multi-blob representation and the experimental one. The simulation data show the sequence of three different stable phase, and compare directly to the experimental results (shown here as a strip)

The whole framework described in the previous sections has been used to predict the phase diagram for diblock copolymers (critical micelle concentration, fluid of micelles, transition between the micellar crystallization and the cylindrical crystalline phases as well as the coexistence between different phases) with an accuracy that allowed to map the theoretical predictions onto experimental results [37, 84, 85, 89] (Fig. 1.11).

Block copolymers are macromolecules characterized by having different regions—blocks—that interact selectively with the solvent they are exposed to. Alternating solvophobic (bad solvent) and solvophillic (good solvent) regions leads to competitions between enthalpic and entropic contributions, hence regions in the chains in which monomers tend to collapse and other regions in which monomers tend to maximize their contact with the solvent. Solutions of block copolymers, as diblock copolymers for example, undergo a macromolecular self-assembly and the subsequent ordering of the self-assembled supermolecular structures.

Exploiting the single chain behavior and building macromolecules with a complex intramolecular architecture opens the path to the design of nano macromolecules, that—if exposed to given physico/chemical conditions—self-assemble into predetermined structures.

The simplest macromolecules that can be assembled out of diblock copolymers are diblock copolymer stars (or telechelic star polymers, TSPs), that are stars that are made of a number f of diblock copolymer arms, made of a percentage α of attractive monomers.

TSPs have been recently introduced as novel self-assembling building blocks ("soft legos"), made of a soft core, decorated by a tunable number of functionalized regions or patches. TSPs have a robust and flexible architecture and they possess the ability to self-assemble at different levels. At the single-molecule level, they first order as soft patchy colloids, which serve then as "soft Legos" for the emergence of larger structures. At the supramolecular level of self-assembly, the soft colloids form complex crystal structures, such as diamond or cubic phases [31].

Grafting solvophilic heads onto a central core, and leaving solvophobic tails exposed to the solvent, leads to the creation of soft and flexible spherical like nanoparticles, decorated by an outer solvophobic shell. Such a shell can either be equally distributed on the surface, or assembled into functionalized regions on the surface. If the density of solvophobic regions is sufficient, the enthalpic contribution becomes an effective attraction between monomers rising from the tendency of the monomers to minimize contacts with the solvent and the solvophobic tails tend to aggregate on the surface of the soft entropic core formed by the solvophobic heads.

The competition between entropic and enthalpic contributions (the choice of the number of arms grafted onto a central core and the percentage of attractive monomers per arm) is the interplay that allows to tunably design particles with functionalized regions: the functionalized regions, in fact, tend to distribute in an ordered way on the surface, giving rise to a soft functionalized colloid.

The number of functionalized regions, or patches, is completely controlled by the choice of a pre-determined set of chemical and/or physical parameters, namely the number f of grafted arms, α and the solvent quality.

A choice of the chemico/physical parameters therefore allows to create fully self-assembling functionalized nanoparticles, whose size and number of patches is completely determined—for fixed solvent quality—by the original (f, α) architecture of the particle. The choice of such two values leads to a single particle phase diagram linking—for each temperature/choice of pH of the solution—the single particle patchiness to the architecture.

For fixed solvent quality, in fact, it is possible to show the soft patchy particle nature of the TSPs and the tunability of the number of patches p as a function of f and α as was shown for small molecules [94, 95] and for large molecules in [31, 96].

It was then shown that such macromolecules had been able to first self-assemble into soft functionalized particle. When at finite density, such particles show the ability of assembling into disordered or ordered structures while retaining the single molecule self-assembled functionalization (Fig. 1.12).

Another interesting path to create tunable building blocks is possible and does not require to change the microscopic nature of the macromolecules in solutions.

Fixed a (f, α) macromolecule architecture, it is possible to exploit the selectivity with which solvophobic tails and solvophilic heads of the stars interact with the solvent, thus controlling the physical and chemical parameters of the solution as a different path to tune the single molecule self-assembly process [97]. Indeed, at fixed external conditions the self-assembly behavior only depends on the number of arms and/or on the ratio of solvophobic to solvophilic monomers. However, changes in temperature and/or solvent quality make it possible to reliably change the number

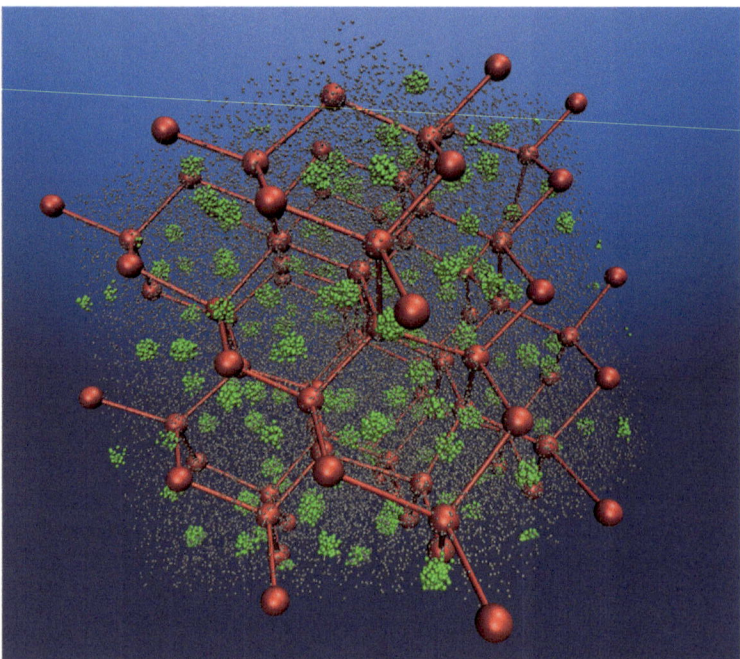

Fig. 1.12 An example of a mechanically stable diamond crystal assembled by TSP with $p = 4$ patches

and size of the attractive patches. This allows to steer the mesoscopic self-assembly behavior without modifying the microscopic constituents.

1.9 Hard-Soft Colloids: Coarse-Graining Strategies for Nanocomposites

The term "nanocomposite" refers to multiphase materials in which one of the phases has a characteristic length scale of less than 100 nm. In this, broadest, sense, the definition can include materials of diverse nature, usually featuring a solid bulk matrix and nano-dimensional phase(s) with dissimilarities in structure and chemistry [98, 99]. In the context of polymeric materials, the field of polymer nanocomposites (PNCs) has been at the very frontier of research in the polymer and soft matter community for the past few decades. While molecular fluids have intermolecular forces fixed by their structural and electronic properties, colloid-polymer systems allow experimentalists to tune these interactions for a wide array of technological applications. PNCs are nowadays a billion-dollar global industry with products ranging from reinforcing components for the transportation industry to commodity plastics with enhanced electrical properties: for an applicative-driven review on the topic, see [99]. From the perspective of Soft Matter physicists, polymer nanocomposites, referred here more loosely as mixture of colloids and polymers at the nanoscale, represent a way to devise a controllable model for fluids. We focus here on binary nanocomposites composed of (1) hard colloids, (2) star polymers, (3) linear chains, as they represent model systems for either uncharged colloidal particles, soft colloids, and polymers in good solvent.

Usually, such polymeric binary nanocomposites are characterized by the *size ratio* $q = R_1/R_2$ where R_1 and R_2 are the characteristic length scales of species 1 and 2, respectively. Two limits can be defined: the "colloid" limit, when colloids (hard or soft) are larger than the polymers (usually, $q < 1$) and the "protein" limit, when the average size of the polymers is larger than the size of the colloids (usually, $q \geq 1$). Such classification is general and crucial in understanding the rich phase diagram of polymer–colloids and polymer–polymer mixtures.

Mixtures of hard colloids and linear chains has been studied in great detail since the first, seminal work of Asakura and Oosawa [100]. From the experimental perspective, depletion forces and phase behavior in such mixtures have been probed by a diverse array of experimental methods, including neutron scattering [101, 102], atomic force microscopy [103], total internal reflection microscopy [104], optical trapping [105], and turbidity measurements [106]. In the colloid limit, several approximate thermodynamic theories have been used, like PRISM approach [107, 108], density-functional [109, 110], thermodynamic perturbation [35, 111], free volume (or osmotic equilibrium) [112, 113], and generalized free volume (GFVT) theories [114, 115]. All these theories, rather accurate in the colloid limit, break down in the protein limit. Full monomer numerical simulations have been attempted [116–118], but they are limited by the large number of degrees of freedom involved. In order to overcome past and present computational limits, several coarse-grained (CG) approaches have been introduced for polymer solutions in different concentration regimes [119–125]: in particular, a multi-blob like approach has been used in the colloid limit [125]. Such approach showed a remarkably favorable comparison with experimental data.

For any value of the size ratio $q = R_g/R_c$, R_g being the radius of gyration of the polymer chain, R_c the radius of the colloid, the system exhibits a liquid–liquid phase separation of the demixing type at low colloid volume fractions. Such transition happens, for increasing q, at higher polymer concentration, but always in the dilute or semi-dilute regime for the polymer chains. At high colloid volume fraction, instead, the addition of a small amount of polymer chains is found to widen the solid–liquid coexistence region [122, 123].

By contrast, star polymer mixtures have become a topic of interest only more recently [126–129]. Star polymers are highly deformable objects that share many features with different hyperbranched polymers including microgels and dendrimers, whose interactions can be easily tuned through several physical and chemical factors [128, 130–132]. In contrast to the classic colloid–polymer example, the exploration of the phase diagram of such mixtures has been tackled only numerically and experimentally. Apart from notable full-monomer simulations [118], a coarse-grained modelization of star-colloid mixtures has been proposed to reliably describe the macroscopic behavior of such mixtures in the full range of size ratios q and star functionalities f [127, 133], i.e. the number of the arms of the stars.

For very low functionality stars and small size ratios $q = R_s/R_c$, R_s being the radius of gyration of the star polymer, stars act as depletants on hard colloids, bringing about depletion effects similar to those seen for colloid-polymers mixtures in the same limit. For high functionalities, the size ratio q has a strong influence on the properties of the binary mixture. While for very small colloidal additive ($q \approx 8$) the glassy behavior of dense star polymer suspensions remains unaffected, mid-sized colloid particles weaken the repulsion between the stars and cause the melting of the glass. Colloidal particles that are even larger ($q \approx 4$) tend to force the system into phase separation. The same generic trends are observed and as f decreases, but higher concentration of colloidal particles can be added before either melting or demixing occurs. This suggests that softer star polymers form more stable mixtures, which can be tuned by adjusting the amount and the size ratio of the colloidal additive. On the other limit, for dense suspensions of hard-spheres the addition of small, non-adsorbing star polymers ($f \gg 1$, $q \approx 0.1$) also leads to a reentrant glass transition. A further increase in polymer concentration leads to an attraction-dominated glass. For larger polymers ($q \approx 0.6$), as the range of the attraction increases, the region of demixing increases in stability relatively to the glass line. Therefore, a range of compositions can be found for which the phase separation can take place without being arrested. The rich phase diagram illustrated above has been confirmed experimentally [134, 135].

The physical mechanism bringing about melting of the glassy state and the subsequent demixing transition can be traced back to depletion. The colloid-induced depletion is superimposed on the soft repulsion between stars and moderate colloid concentration causes a reduction in the star–star repulsion, leading to melting of the glass. On the other limit, at sufficiently high densities, the pure colloidal system also reaches a repulsive glassy state and, once a small amount of polymer is added, the induced depletion potential makes the particles cluster, restoring the ergodicity [134–136].

Finally, even less attention has been devoted to star polymer–linear polymeric chain nanocomposites. In the colloid limit, $q = R_g/R_s < 1$, a successful modelization has been established [137–143] employing a coarse-graining strategy in which the whole polymer chain was represented by a single degree of freedom (single *blob*), typically its central monomer. Such coarse-graining has been successfully validated against experimental data and has been able to rationalize the phase diagram of star-chain mixtures which, in the colloid limit, features the melting of the glass of the stars and the (arrested) phase separation upon further addition of linear chains [141]. In the protein limit, this coarse-graining loses its validity fundamentally, not just quantitatively [123], as which drastic departures of the polymer conformation from an average, "soft sphere" shape are expected. A multi-blob coarse-graining of the linear chain in such nanocomposite has been introduced [144], employing a regrouping of degrees of freedom that allows for a simplification of the system while retaining the ability of reproducing its essential features as a polymer chain. At the coarse-grained level, the long chain is replaced by a succession of n effective monomers, or *blobs*, each of which represents a small sub-part of the M-long polymeric chain, each blob made of $m = M/n$ monomers. Every blob sees the star as a single, coarse-grained object. The star–blob interaction $V_{sb}(r)$ is obtained by computing, in the zero-density limit, the effective potential between a star polymer and a short polymer chain of length $m = 10$ monomers. This blob-based approach permits a realistic description of long polymeric chains, accessing information on the conformational properties of the latter while reducing the star to a single interaction center, hence achieving a huge computational gain with respect to a full monomer resolved simulation.

Within the linear chain, each blob is connected to its neighbors by a harmonic potential

$$\beta V_{\text{conn}}(s) = \frac{\kappa}{2}(s - s_0)^2, \tag{1.48}$$

where s is the separation between the centers of mass of two (connected) blobs, $\kappa = 0.11$ and $s_0 = R_b$, R_b being the radius of gyration of a polymer chain of length m. Each blob interacts with any other blob, connected to it or not, through a Gaussian steric potential [145]

$$\beta V_{\text{steric}}(s) = A \exp\left[-\alpha (s/s_0)^2\right], \tag{1.49}$$

where $\alpha = 1.98, A = 2.45$.

Notice that the intramolecular potentials used to describe the chain are extracted in proximity of a star polymer; in such a way both the intramolecular and the chain-star many-body contribution are included in the effective interactions. The fact that blobs only contain a few monomers each (hence the subsegments of the chain are off-scaling) and the influence on the effective representation due to the presence of the star leads to a discrepancy between the here used potentials and the ones known to well describe homopolymeric chains [145]; effective potentials have in fact, in this context, a shorter ranged blob–blob interaction and a weaker tethering. Such a discrepacy arises from the stretching that the chain acts when the center of mass of the homopolymer approaches the center of the star. To mimic such a conformational property, the intramolecular effective potentials computed within the star are softer than expected.

The natural choice of effective coordinates to describe the interaction between the star and the short polymer blob are the central anchoring point of the former and the center of mass of the latter. Accordingly, the star-blob effective potential $V_{sb}(r)$, where r is the distance between the two effective coordinates, has been computed employing the Widom insertion method [146], which guarantees excellent precision at both short and large distances, being at the same time computationally inexpensive. The star–blob interaction has been found to possess an interesting scaling property, as a universal function of $r/(\sigma \sqrt{f})$ can be defined, for relatively small values of $r/(\sigma \sqrt{f})$, roughly up to $R_{sb}/(\sigma \sqrt{f})$. Here $R_{sb} = R_s + R_b$, R_s being the gyration radius of the star and $R_b = 1.548\,\sigma$ being the radius of gyration of the $m = 10$-short chain. Following the Daoud and Cotton representation of a star polymer [147], for distances $r < R_s$ from its center, the star can be modeled as a succession of concentric shell of Daoud-Cotton (DC) blobs of size $\xi(r)$; moreover, for $r < R_s$ the local correlation length $\xi(r)$ increases as $\xi(r) \propto r/(\sigma\sqrt{f})$. Within the star interior, an inhomogeneous semi-dilute polymer solution is formed and the DC blob is the local correlation length $\xi(r)$: accordingly [148], the potential between the short chain and the semidilute-solution DC blobs is the same for DC blobs of the same size $\xi(r)$. Therefore, for two stars of different functionalities, f_1 and f_2, the free energy cost of insertion of the short chain fulfills the property

$$V_{sb}(r,f_1) = V_{sb}\left(\frac{r}{\sigma}\sqrt{\frac{f_2}{f_1}},f_2\right) \tag{1.50}$$

as DC blob having size $\xi(r)$ in a star of functionality f_1 will be found at a distance $r\sqrt{f_2/f_1}$ in a star of functionality f_2. This implies a scaling property with r and f, namely

$$V_{sb}(r,f) = \phi(r/(\sigma\sqrt{f})), \qquad (r \lesssim R_s) \tag{1.51}$$

This argument is valid as long as the object interacting with the star polymer is sufficiently small with respect to R_s. For $r > R_{sb}$, the potential is found to be a function of the cross-radius of gyration R_{sb} only. These features lead to a general prediction of the shape of the interaction potential for arbitrary f and can be exploited to obtain an approximate, analytical form of the star–blob interaction potential

$$\beta V_{sb}(r) = \begin{cases} v_1\left(\frac{r}{\sqrt{f}}\right)^{-b_1} + c_1; & \text{for } r/\sigma \leq \sqrt{f/e}, \\ v_2\left(\frac{r}{\sqrt{f}}\right)^{-b_2}; & \text{for } \sqrt{f/e} < r/\sigma \leq R_{sb}, \\ v_3\exp\left[-(r-R_{sb})^2/(wR_{sb}^2)\right]; & \text{for } R_{sb} < r/\sigma, \end{cases} \tag{1.52}$$

where $v_1 = 0.0477$, $b_1 = 8.1279$, $c_1 = 9.0306$, $v_2 = 3.7973$, $b_2 = 2.091$, have been obtained through a fitting procedure, bound to ensure that the potential is continuous at $r/(\sigma\sqrt{f}) = 1/\sqrt{e}$. The parameter v_3 has to be chosen such that the potential is continuous at $r = R_{sb}$. Last, the parameter w, which regulates the width of the Gaussian, is obtained again by a fitting procedure. For stars with longer arms, fixing $w = 0.145$ yields consistently an excellent agreement with the numerical data.

Confirmation of the validity of the multi-blob coarse-grained approach comes from the striking comparison between effective potentials between stars and long chains, obtained within the two representations; moreover, the drastic reduction of the degrees of freedom grants an impressive computational gain. Additional, strong corroboration of the result comes from the analysis of the shape and of the orientations of the chain in both representations. The shape of the chain has been analyzed by computing the gyration tensor. In both representations, the largest eigenvalue is maximal when the separation between the center of the two objects is small and decreases until it reaches a constant value for increasing separations. Within both representation, the chain, while in proximity of the star, stretches to adapt to its complex structure.

Finally, the relative orientations, denoted by θ_i, between every eigenvector $\hat{e}_i$ of the gyration tensor and the vector $\vec{r}$ connecting the star and chain centers has been analyzed. In both cases, the directions of the major and minor axis of the polymer are favored to be perpendicular to the vector connecting the star and chain centers, whereas the other one is favored to be parallel: the chain is trying to stretch and to fit within the star at short separations, attempting at the same time to "embrace" it, as much as possible. At large separations, all distributions become flat, as expected. This analysis shows that the multi-blob-based approach is, as a matter of fact, able to reproduce quite faithfully the conformational properties of the chain. Application of such multi-blob coarse-graining on the phase diagram of star-chain nanocomposites will be subject of future investigation.

1.10 Concluding Remarks

We have presented a detailed exposure to a multiscale strategy to coarse-grain polymeric systems in the semi-dilute regime, starting from the single molecule details up to a deep exploration of the semi-dilute regime; the simplification in the description and the controlled reduction of the degrees of freedom allows, on the one hand, to explore by means of numerical simulations a region of the phase space where many-body interactions play a crucial role, and on the other hand to retain all of the original information on the microscopic system thus being able to map results and theoretical predictions on experimental realisations.

Same methodology and approach can be used when different species are considered, as in the case of nanocomposites, where different species and different scales are to be considered.

Acknowledgements B. C. acknowledges support from the Austrian Academy of Sciences (ÖAW) through her APART Fellowship No. 11723.

References

1. Maldovan M, Thomas EL. Diamond-structured photonic crystals. Nat Mater. 2004;3:593.
2. Damasceno PF, Engel M, Glotzer SC. Predictive self-assembly of polyhedra into complex structures. Science. 2012;337(6093):453–7.
3. Glotzer SC, Solomon MJ. Anisotropy of building blocks and their assembly into complex structures. Nat Mater. 2007;6(8):557–62.
4. Torquato S, Jiao Y. Dense packings of the Platonic and Archimedean solids. Nature. 2009;460(7257):876–9.

5. Akcora P, Liu H, Kumar SK, Moll J, Li Y, Benicewicz BC, Schadler LS, Acehan D, Panagiotopoulos AZ, Pryamitsyn V, et al. Anisotropic self-assembly of spherical polymer-grafted nanoparticles. Nat Mater. 2009;8(4):354–9.
6. Bianchi E, Blaak R, Likos CN. Patchy colloids: state of the art and perspectives. Phys Chem Chem Phys. 2011;13(14):6397–410.
7. Ferrari S, Bianchi E, Kalyuzhnyi YV, Kahl G. Inverse patchy colloids with small patches: fluid structure and dynamical slowing down. J Phys Condens Matter. 2015;27(23):234104.
8. Bianchi E, Kahl G, Likos CN, Sciortino F. Patchy particles. J Phys Condens Matter. 2015;27:230301.
9. van Oostrum PDJ, Hejazifar M, Niedermayer C, Reimhult E. Simple method for the synthesis of inverse patchy colloids. J Phys Condens Matter. 2015;27(23):234105.
10. Grünewald TA, Lassenberger A, van Oostrum PDJ, Rennhofer H, Zirbs R, Capone B, Vonderhaid I, Amenitsch H, Lichtenegger HC, Reimhult E. Core–shell structure of monodisperse poly(ethylene glycol)-grafted iron oxide nanoparticles studied by small-angle X-ray scattering. Chem Mater. 2015;27(13):4763–71.
11. Choueiri RM, Galati E, Thérien-Aubin H, Klinkova A, Larin EM, Querejeta-Fernández A, Han L, Xin HL, Gang O, Zhulina EB, Rubinstein M, Kumacheva E. Surface patterning of nanoparticles with polymer patches. Nature. 2016;538(7623):79–83.
12. Romano F, Sanz E, Sciortino F. Phase diagram of a tetrahedral patchy particle model for different interaction ranges. J Chem Phys. 2010;132(18):184501.
13. Snyder CE, Yake AM, Feick JD, Velegol D. Nanoscale functionalization and site-specific assembly of colloids by particle lithography. Langmuir. 2005;21(11):4813–5.
14. Yake AM, Snyder CE, Velegol D. Site-specific functionalization on individual colloids: size control, stability, and multilayers. Langmuir. 2007;23(17):9069–75.
15. Nie Z, Li W, Seo M, Xu S, Kumacheva E. Janus and ternary particles generated by microfluidic synthesis: design, synthesis, and self-assembly. J Am Chem Soc. 2006;128(29):9408–12.
16. Pawar AB, Kretzschmar I. Patchy particles by glancing angle deposition. Langmuir. 2008;24(2):355–8.
17. Pawar AB, Kretzschmar I. Multifunctional patchy particles by glancing angle deposition. Langmuir. 2009;25(16):9057–63.
18. Mognetti BM, Leunissen ME, Frenkel D. Controlling the temperature sensitivity of DNA-mediated colloidal interactions through competing linkages. Soft Matter. 2012;8(7):2213–21.
19. Angioletti-Uberti S, Mognetti BM, Frenkel D. Re-entrant melting as a design principle for DNA-coated colloids. Nat Mater. 2012;11(6):518–22.
20. Mayer AC, Scully SR, Hardin BE, Rowell MW, McGehee MD. Polymer-based solar cells. Mater Today. 2007;10(11):28–33.
21. de Gennes PG. Scaling concepts in polymer physics. Ithaca: Cornell University Press; 1979.
22. Flory PJ, Krigbaum WR. Statistical mechanics of dilute polymer solutions. II. J Chem Phys. 1950;18(8):1086–94.
23. Flory PJ. Principles of polymer chemistry. Itacha: Cornell University Press; 1953.
24. Doi M, Edwards SF. The theory of polymer dynamics. Oxford: Oxford University Press; 1986.
25. Doi M. Introduction to polymer physics. Oxford: Clarendon Press; 1996.
26. des Cloizeaux J, Jannink G. Polymers in solution, their modelling and structure. Oxford: Oxford University Press; 1990.
27. Pelissetto A, Hansen J-P. Corrections to scaling and crossover from good-to θ-solvent regimes of interacting polymers. J Chem Phys. 2005;122(13):134904.
28. Bolhuis PG, Louis AA, Hansen J-P, Meijer EJ. J Chem Phys. 2001;114(9):4296.
29. Henderson RL. A uniqueness theorem for fluid pair correlation functions. Phys Lett A. 1974;49(3):197–8.
30. Hansen J-P, Mc Donald IR. Theory of simple liquids. 3rd ed. Amsterdam: Academic; 2006.
31. Capone B, Coluzza I, LoVerso F, Likos CN, Blaak R. Telechelic star polymers as self-assembling units from the molecular to the macroscopic scale. Phys Rev Lett. 2012;109(23):238301.

32. Pierleoni C, Capone B, Hansen J-P. J Chem Phys. 2007;127(17):171102.
33. Grosberg AY, Khalatur PG, Khokhlov AR. Polymeric coils with excluded volume in dilute solution: the invalidity of the model of impenetrable spheres and the influence of excluded volume on the rates of diffusion-controlled intermacromolecular reactions. Makromol Chem Rapid. 1982;3(10):709–13.
34. Dautenhahn J, Hall CK. Macromolecules. 1994;27:5399.
35. Pelissetto A, Hansen J-P. An effective two-component description of colloid-polymer phase separation. Macromolecules. 2006;39(26):9571–80.
36. Pierleoni C, Capone B, Hansen J-P. A soft effective segment representation of semidilute polymer solutions. J Chem Phys. 2007;127(17):171102.
37. Capone B, Hansen JP, Coluzza I. Competing micellar and cylindrical phases in semi-dilute diblock copolymer solutions. Soft Matter. 2010;6:6075.
38. Capone B, Coluzza I, Hansen J-P. A systematic coarse-graining strategy for semi-dilute copolymer solutions: from monomers to micelles. J Phys Condens Matter. 2011;23:194102.
39. Coluzza I, Capone B, Hansen J-P. Rescaling of structural length scales for "soft effective segment" representations of polymers in good solvent. Soft Matter. 2011;7:5255.
40. Ladanyi BM, Chandler D. New type of cluster theory for molecular fluids: interaction site cluster expansion. J Chem Phys. 1975;62(11):4308–24.
41. Louis AA. Beware of density dependent pair potentials. J Phys Condens Matter. 2002;14(40):9187.
42. Micheletti C, Marenduzzo D, Orlandini E. Polymers with spatial or topological constraints: theoretical and computational results. Phys Rep. 2011;504(1):1–73.
43. Müller M, Wittmer JP, Cates ME. Topological effects in ring polymers: a computer simulation study. Phys Rev E. 1996;53(5):5063.
44. Grosberg AY. Annealed lattice animal model and Flory theory for the melt of non-concatenated rings: towards the physics of crumpling. Soft Matter. 2014;10:560–5.
45. Takano A, Kushida Y, Ohta Y, Masuoka K, Matsushita Y. The second virial coefficients of highly-purified ring polystyrenes in cyclohexane. Polymer. 2009;50:1300.
46. Narros A, Moreno AJ, Likos CN. Effects of knots on ring polymers in solvents of varying quality. Macromolecules. 2013;46:3654.
47. Hirayama N, Tsurusaki K, Deguchi T. Linking probabilities of off-lattice self-avoiding polygons and the effects of excluded volume. J Phys A: Math Theor. 2009;42:05001.
48. Bernabei M, Bacova P, Moreno AJ, Narros A, Likos CN. Fluids of semiflexible ring polymers: effective potentials and clustering. Soft Matter. 2013;9(4):1287–300.
49. Rubinstein M, Colby RH. Polymer physics. Oxford: Oxford University Press; 2003.
50. Roovers J. The melt properties of ring polystyrenes. Macromolecules. 1985;18:1359.
51. Roovers J. Viscoelastic properties of polybutadiene rings. Macromolecules. 1988;21(5):1517–21.
52. McKenna GB, Hadziioannou G, Lutz P, Hild G, Strazielle C, Straupe C, Rempp P, Kovacs AJ. Dilute solution characterization of cyclic polystyrene molecules and their zero-shear viscosity in the melt. Macromolecules. 1987;20(3):498–512.
53. Hodgson DF, Amis EJ. Dilute solution behavior of cyclic and linear polyelectrolytes. J Chem Phys. 1991;95:7653.
54. Brown S, Szamel G. Structure and dynamics of ring polymers. J Chem Phys. 1998;108(12):4705–8.
55. Brown S, Szamel G. Computer simulation study of the structure and dynamics of ring polymers. J Chem Phys. 1998;109(14):6184–92.
56. Brown S, Lenczycki T, Szamel G. Influence of topological constraints on the statics and dynamics of ring polymers. Phys Rev E 2001;63(5):052801.
57. Müller M, Wittmer JP, Cates ME. Topological effects in ring polymers. II. Influence of persistence length. Phys Rev E. 2000;61(4):4078.
58. Hur K, Jeong C, Winkler RG, Lacevic N, Gee RH, Yoon DY. Chain dynamics of ring and linear polyethylene melts from molecular dynamics simulations. Macromolecules. 2011;44(7):2311–5.

59. Halverson JD, Lee WB, Grest GS, Grosberg AY, Kremer K. Molecular dynamics simulation study of nonconcatenated ring polymers in a melt. II. Dynamics. J Chem Phys. 2011;134(20):204905.

60. Tsolou G, Stratikis N, Baig C, Stephanou PS, Mavrantzas VG. Melt structure and dynamics of unentangled polyethylene rings: rouse theory, atomistic molecular dynamics simulation, and comparison with the linear analogues. Macromolecules. 2010;43(24):10692–713.

61. Kapnistos M, Lang M, Vlassopoulos D, Pyckhout-Hintzen W, Richter D, Cho D, Chang T, Rubinstein M. Unexpected power-law stress relaxation of entangled ring polymers. Nat Mater. 2008;7(12):997–1002.

62. Slimani MZ, Bacova P, Bernabei M, Narros A, Likos CN, Moreno AJ. Cluster glasses of semiflexible ring polymers. ACS Macro Lett. 2014;3(7):611–6.

63. Cates ME, Deutsch JM. Conjectures on the statistics of ring polymers. J Phys Fr. 1986;47(12):2121.

64. Pakula T, Geyler S. Cooperative relaxations in condensed macromolecular systems. 3. Computer-simulated melts of cyclic polymers. Macromolecules. 1988;21(6):1665–70.

65. Müller M, Wittmer JP, Barrat J-L. On two intrinsic length scales in polymer physics: topological constraints vs. entanglement length. Europhys Lett. 2000;52(4):406.

66. Hur K, Winkler RG, Yoon DY, et al. Comparison of ring and linear polyethylene from molecular dynamics simulations. Macromolecules. 2006;39(12):3975–7.

67. Vettorel T, Grosberg AY, Kremer K. Statistics of polymer rings in the melt: a numerical simulation study. Phys Biol. 2009;6(2):025013.

68. Suzuki J, Takano A, Deguchi T, Matsushita Y. Dimension of ring polymers in bulk studied by Monte-Carlo simulation and self-consistent theory. J Chem Phys. 2009;131(14):144902.

69. Halverson JD, Lee WB, Grest GS, Grosberg AY, Kremer K. Molecular dynamics simulation study of nonconcatenated ring polymers in a melt. I. Statics. J Chem Phys. 2011;134(20):204904.

70. Sakaue T. Ring polymers in melts and solutions: scaling and crossover. Phys Rev Lett. 2011;106(16):167802.

71. Sakaue T. Statistics and geometrical picture of ring polymer melts and solutions. Phys Rev E. 2012;85(2):021806.

72. Everaers R, Sukumaran SK, Grest GS, Svaneborg C, Sivasubramanian A, Kremer K. Rheology and microscopic topology of entangled polymeric liquids. Science. 2004;303(5659):823–6.

73. Narros A, Moreno AJ, Likos CN. Influence of topology on effective potentials: coarse-graining ring polymers. Soft Matter. 2011;6:2435.

74. Safran SA. Statistical thermodynamics of surfaces, interfaces and membranes. Reading: Addison Wesley; 1994.

75. Hamley IW. Block copolymers in solution. Chichester: Wiley; 2005.

76. Leibler L. Theory of microphase separation in block copolymers. Macromolecules. 1980;13(6):1602–17.

77. Matsen MW, Schick M. Stable and unstable phases of a diblock copolymer melt. Phys Rev Lett. 1994;72(16):2660.

78. Ohta T, Kawasaki K. Equilibrium morphology of block copolymer melts. Macromolecules. 1986;19(10):2621–32.

79. Fredrickson GH, Helfand E. Fluctuation effects in the theory of microphase separation in block copolymers. J Chem Phys. 1987;87(1):697–705.

80. Zhulina EB, Adam M, LaRue I, Sheiko SS, Rubinstein M, et al. Diblock copolymer micelles in a dilute solution. Macromolecules. 2005;38(12):5330–51.

81. Milchev A, Bhattacharya A, Binder K. Formation of block copolymer micelles in solution: a Monte Carlo study of chain length dependence. Macromolecules. 2001;34(6):1881–93.

82. Termonia Y. Sphere-to-cylinder transition in dilute solutions of diblock copolymers. J Polym Sci B Polym Phys. 2002;40(9):890–5.

83. Wijmans CM, Eiser E, Frenkel D. Simulation study of intra-and intermicellar ordering in triblock-copolymer systems. J Chem Phys. 2004;120(12):5839–48.

84. Lodge TL, Bang J, Li Z, Hillmyer MA, Talmon Y. Strategies for controlling intra-and intermicellar packing in block copolymer solutions: illustrating the flexibility of the self-assembly toolbox. Faraday Discuss. 2004;128:1–12.
85. Lodge TP, Pudil B, Hanley KJ. The full phase behavior for block copolymers in solvents of varying selectivity. Macromolecules. 2002;35(12):4707–17.
86. Park MJ, Char K, Bang J, Lodge TP. Interplay between cubic and hexagonal phases in block copolymer solutions. Langmuir. 2005;21(4):1403–11.
87. Addison CI, Hansen JP, Krakoviack V, Louis AA. Coarse-graining diblock copolymer solutions: a macromolecular version of the Widom–Rowlinson model. Mol Phys. 2005;103(21–23):3045–54.
88. Bang J, Lodge TP. On the selection of FCC and BCC lattices in poly (styrene-b-isoprene) copolymer micelles. Macromol Res. 2008;16(1):51–6.
89. McConnell GA, Gast AP. Melting of ordered arrays and shape transitions in highly concentrated diblock copolymer solutions. Macromolecules. 1997;30(3):435–44.
90. Ziherl P, Kamien RD. Soap froths and crystal structures. Phys Rev Lett. 2000;85(16):3528.
91. Molina JJ, Pierleoni C, Capone B, Hansen J-P, Saulo Santos de Oliveira I. Crystal stability of diblock copolymer micelles in solution. Mol Phys. 2009;107(4–6):535–48.
92. Widom B. Some topics in theory of fluids. J Chem Phys. 1963;39(11):2808.
93. Bates FS, Fredrickson GH. Block copolymer thermodynamics: theory and experiment. Annu Rev Phys Chem. 1990;41(1):525–57.
94. Lo Verso F, Likos CN, Mayer C, Löwen H. Collapse of telechelic star polymers to watermelon structures. Phys Rev Lett. 2006;96:187802.
95. Lo Verso F, Panagiotopoulos AZ, Likos CN. Aggregation phenomena in telechelic star polymer solutions. Phys Rev E. 2009;79:010401.
96. Capone B, Coluzza I, Blaak R, Lo Verso F, Likos CN. Hierarchical self-assembly of telechelic star polymers: from soft patchy particles to gels and diamond crystals. J Phys. 2013;15(9):095002.
97. Rovigatti L, Capone B, Likos CN. Soft self-assembled nanoparticles with temperature-dependent properties. Nanoscale. 2016;8(6):3288–95.
98. Kumar SK, Krishnamoorti R. Nanocomposites: structure, phase behavior, and properties. Annu Rev Chem Biomol Eng. 2010;1(1):37–58.
99. Kumar SK, Benicewicz BC, Vaia RA, Winey KI. 50th anniversary perspective: are polymer nanocomposites practical for applications? Macromolecules. 2017;50(3):714–31.
100. Asakura S, Oosawa F. On interaction between two bodies immersed in a solution of macromolecules. J Chem Phys. 1954;22:1255–6.
101. Le Coeur C, Teixeira J, Busch P, Longeville S. Compression of random coils due to macromolecular crowding: scaling effects. Phys Rev E. 2010;81:061914.
102. Nusser K, Neueder S, Schneider GJ, Meyer M, Pyckhout-Hintzen W, Willner L, Radulescu A, Richter D. Conformations of SilicaPoly(ethylenepropylene) nanocomposites. Macromolecules. 2010;43(23):9837–47.
103. Milling A, Biggs S. Direct measurement of the depletion force using an atomic force microscope. J Colloid Interface Sci. 1995;170(2):604–6.
104. Rudhardt D, Bechinger C, Leiderer P. Direct measurement of depletion potentials in mixtures of colloids and nonionic polymers. Phys Rev Lett. 1998;81:1330–3.
105. Hilitski F, Ward AR, Cajamarca L, Hagan MF, Grason GM, Dogic Z. Measuring cohesion between macromolecular filaments one pair at a time: depletion-induced microtubule bundling. Phys Rev Lett. 2015;114:138102.
106. Hennequin Y, Evens M, Quezada Angulo CM, van Duijneveldt JS. Miscibility of small colloidal spheres with large polymers in good solvent. J Chem Phys. 2005;123(5):054906.
107. Fuchs M, Schweizer KS. Structure and thermodynamics of colloid-polymer mixtures: a macromolecular approach. Europhys Lett. 2000;51(6):621.
108. Fuchs M, Schweizer KS. Structure of colloid-polymer suspensions. J Phys Condens Matter. 2002;14(12):R239.

109. Schmidt M, Denton AR, Brader JM. Fluid demixing in colloidpolymer mixtures: influence of polymer interactions. J Chem Phys. 2003;118(3):1541–9.
110. Denton AR, Schmidt M. Colloid-induced polymer compression. J Phys Condens Matter. 2002;14(46):12051.
111. Paricaud P, Varga S, Jackson G. Study of the demixing transition in model athermal mixtures of colloids and flexible self-excluding polymers using the thermodynamic perturbation theory of Wertheim. J Chem Phys. 2003;118(18):8525–36.
112. Lekkerkerker HNW, Poon WC-K, Pusey PN, Stroobants A, Warren PB. Phase behaviour of colloid + polymer mixtures. Europhys. Lett. 1992;20(6):559.
113. Lekkerkerker HNW. Osmotic equilibrium treatment of the phase separation in colloidal dispersions containing non-adsorbing polymer molecules. Colloids Surf. 1990;51:419–26.
114. Fleer GJ, Tuinier R. Analytical phase diagram for colloid-polymer mixtures. Phys Rev E. 2007;76:041802.
115. Fleer GJ, Tuinier R. Analytical phase diagrams for colloids and non-adsorbing polymer. Adv Colloid Interf Sci. 2008;143(1–2):1–47.
116. Chou C-Y, Vo TTM, Panagiotopoulos AZ, Robert M. Computer simulations of phase transitions of bulk and confined colloid-polymer systems. Physica A. 2006;369(2):275–90.
117. Mahynski NA, Lafitte T, Panagiotopoulos AZ. Pressure and density scaling for colloid-polymer systems in the protein limit. Phys Rev E. 2012;85(5):051402.
118. Mahynski NA, Panagiotopoulos AZ. Phase behavior of athermal colloid-star polymer mixtures. J Chem Phys. 2013;139(2):024907.
119. Dijkstra M, Brader JM, Evans R. Phase behaviour and structure of model colloid-polymer mixtures. J Phys Condens Matter. 1999;11(50):10079.
120. Louis AA, Bolhuis PG, Hansen JP, Meijer EJ. Can polymer coils be modeled as "Soft Colloids"? Phys Rev Lett. 2000;85:2522–25.
121. Bolhuis PG, Louis AA, Hansen JP, Meijer EJ. Accurate effective pair potentials for polymer solutions. J Chem Phys. 2001;114(9):4296–311.
122. Bolhuis PG, Louis AA, Hansen J-P. Influence of polymer-excluded volume on the phase-behavior of colloid-polymer mixtures. Phys Rev Lett. 2002;89:128302.
123. Bolhuis PG, Meijer EJ, Louis AA. Colloid-polymer mixtures in the protein limit. Phys Rev Lett. 2003;90:068304.
124. Lim WK, Denton AR. Depletion-induced forces and crowding in polymer-nanoparticle mixtures: role of polymer shape fluctuations and penetrability. J Chem Phys. 2016;144(2):024904.
125. D'Adamo G, Pelissetto A, Pierleoni C. Phase diagram and structure of mixtures of large colloids and linear polymers under good-solvent conditions. Macromolecules. 2016;49(14):5266–80.
126. von Ferber C, Jusufi A, Watzlawek M, Likos CN, Löwen H. Polydisperse star polymer solutions. Phys Rev E. 2000;62:6949–56.
127. Jusufi A, Dzubiella J, Likos CN, von Ferber C, Lwen H. Effective interactions between star polymers and colloidal particles. J Phys Condens Matter. 2001;13(28):6177.
128. Likos CN. Soft matter with soft particles. Soft Matter. 2006;2:478–98.
129. Mayer C, Zaccarelli E, Stiakakis E, Likos CN, Sciortino F, Munam A, Gauthier M, Hadjichristidis N, Iatrou H, Tartaglia P, et al. Asymmetric caging in soft colloidal mixtures. Nat Mater. 2008;7(10):780–4.
130. Grest GS, Fetters LJ, Huang JS, Richter D. Star polymers: experiment, theory, and simulation. Adv Chem Phys. 1996;94:67.
131. Likos CN, Löwen H, Watzlawek M, Abbas B, Jucknischke O, Allgaier J, Richter D. Star polymers viewed as ultrasoft colloidal particles. Phys Rev Lett. 1998;80:4450.
132. Likos CN. Effective interactions in Soft Condensed Matter Physics. Phys Rep. 2001;348:267–439.
133. Marzi D, Likos CN, Capone B. Coarse graining of star-polymer colloid nanocomposites. J Chem Phys. 2012;137(1):014902.

134. Truzzolillo D, Marzi D, Marakis J, Capone B, Camargo M, Munam A, Moingeon F, Gauthier M, Likos CN, Vlassopoulos D. Glassy states in asymmetric mixtures of soft and hard colloids. Phys Rev Lett. 2013;111:208301.
135. Marzi D, Capone B, Marakis J, Merola MC, Truzzolillo D, Cipelletti L, Moingeon F, Gauthier M, Vlassopoulos D, Likos CN, Camargo M. Depletion, melting and reentrant solidification in mixtures of soft and hard colloids. Soft Matter. 2015;11:8296.
136. Dzubiella J, Jusufi A, Likos CN, von Ferber C, Löwen H, Stellbrink J, Allgaier J, Richter D, Schofield AB, Smith PA, Poon WCK, Pusey PN. Phase separation in star polymer-colloid mixtures. Phys Rev E. 2001;64:010401.
137. Camargo M, Likos CN. Phase separation in star-linear polymer mixtures. J Chem Phys. 2009;130:204904.
138. Camargo M, Likos CN. Unusual features of depletion interactions in soft polymer-based colloids mixed with linear homopolymers. Phys Rev Lett. 2010;104:078301.
139. Stiakakis E, Vlassopoulos D, Likos CN, Roovers J, Meier G. Polymer-mediated melting in ultrasoft colloidal gels. Phys Rev Lett. 2002;89:208302.
140. Stiakakis E, Petekidis G, Vlassopoulos D, Likos CN, Iatrou H, Hadjichristidis N, Roovers J. Depletion and cluster formation in soft colloid - polymer mixtures. Europhys Lett. 2005;72:664–70.
141. Lonetti B, Camargo M, Stellbrink J, Likos CN, Zaccarelli E, Willner L, Lindner P, Richter D. Ultrasoft colloid-polymer mixtures: structure and phase diagram. Phys Rev Lett. 2011;106:228301.
142. Truzzolillo D, Vlassopoulos D, Gauthier M. Osmotic interactions, rheology, and arrested phase separation of star-linear polymer mixtures. Macromolecules. 2011;44:5043.
143. Wilk A, Huißmann S, Stiakakis E, Kohlbrecher J, Vlassopoulos D, Likos CN, Meier G, Dhont JKG, Petekidis G, Vavrin R. Osmotic shrinkage in star/linear polymer mixtures. Eur Phys J E. 2010;32:127–34.
144. Locatelli E, Capone B, Likos CN. Multiblob coarse-graining for mixtures of long polymers and soft colloids. J Chem Phys. 2016;145(17):174901.
145. Pierleoni C, Capone B, Hansen JP. A soft effective segment representation of semidiluite polymer solutions. J Chem Phys. 2007;127:171102.
146. Mladek BM, Frenkel D. Pair interactions between complex mesoscopic particles from Widom particle-insertion method. Soft Matter. 2011;7:1450–5.
147. Daoud M, Cotton JP. Star shaped polymers: a model for the conformation and its concentration dependence. J Phys. 1982;43:531.
148. Sear RP. Entropy-driven phase separation in mixtures of small colloidal particles and semidilute polymers. Phys Rev E. 1997;56:4463.

Modeling the Effective Interactions Between Heterogeneously Charged Colloids to Design Responsive Self-assembled Materials

2

Emanuela Bianchi

2.1 Introduction

Within the framework of rational materials design, colloids with a non-homogeneous surface charge distribution have started to attract the interest of the scientific community in recent years [1–8]. The key advantage of heterogeneously charged units at the nano- and micro-scale is their ability to combine directional bonding with a simple particle design. Thanks to the competitive interplay between orientation-dependent attraction and repulsion—induced by the interactions between like and oppositely charged regions on the particle surface—experimentally accessible surface patterns are complex enough to favor the stabilization of desired target structures. Moreover, due to the electrostatic nature of the inter-particle interactions, heterogeneously charged units are characterized by a tunable responsiveness to external stimuli, such as variations of pH and changes of salt concentration, allowing for a fine control over the properties of the desired phases, either before, during or after the assembly.

Heterogeneously charged particles can be generally regarded as charged patchy colloids. Conventional patchy particles are colloids with chemically or physically patterned surfaces that are able to guarantee a fine control over the symmetries of the equilibrium phases: the orientational and possibly selective bonding mechanism mediated by the patches determines the symmetries of the resulting structures [9–12] (for an overview on the synthesis of patchy colloids, we refer to Chap. 4). To emphasize that heterogeneously charged units are a different class of patchy colloids with respect to conventional patchy particles, they are often named inverse patchy colloids (IPCs): the term inverse refers to the fact that, while conventional patchy

E. Bianchi (✉)
Faculty of Physics, University of Vienna, Boltzmanngasse 5, A-1090 Vienna, Austria

Institut für Theoretische Physik, TU Wien, Wiedner Hauptstraße 8-10, A-1040 Vienna, Austria
e-mail: emanuela.bianchi@univie.ac.at

© Springer International Publishing AG, part of Springer Nature 2017
I. Coluzza (ed.), *Design of Self-Assembling Materials*,
https://doi.org/10.1007/978-3-319-71578-0_2

systems are typically characterized by the presence of attractive regions on the surface of otherwise repulsive particles, IPCs carry extended patches that repel each other and attract those parts of the colloid that are free of patches. The effective interaction between IPCs can be both attractive and repulsive, depending on the relative orientation of the particles. This class of systems was originally introduced to describe complex units emerging from the adsorption of charged polyelectrolyte stars onto the surface of oppositely charged colloids [13], but it has soon included one-component systems of heterogeneously charged, micro-scale particles [2]. The first elaborate model, for which the term IPCs was coined, was put forward about 6 years ago in [1] for colloidal particles with two charged polar patches and an oppositely charged equatorial belt. In the subsequent years, the idea to study particles with heterogeneously charged surfaces proliferated within the community: models of charged Janus particles have been proposed in [6] and [8]; a set of charged patchy particle models was introduced in [4] within the context of globular proteins; zwitterionic and protein-like units have also been modeled in [5]; finally, a highly sophisticated model for particles with heterogeneous icosahedral, octahedral, and tetrahedral charge decorations has been put forward in [14] to describe virus capsids.

Within this broad class of systems, most of the attention has been devoted so far to IPCs with two charged polar patches and an oppositely charged equatorial belt. Recently, a novel method that does not rely on clean-room facilities and that is easily scalable has allowed to modify the surface of colloidal particles by creating two polar regions and an oppositely charged equatorial region, the net charge of the resulting particles being controlled via the pH of the solution [2]. Numerical investigations based on a suitably developed coarse-grained model for IPCs with two identical polar patches have shown that an emerging feature of these systems is the formation of planar aggregates either as monolayers close to a possibly charged substrate or as bulk equilibrium phases [15–19]. Close to a homogeneously charged substrate, IPCs with two identical patches form complex structures with well-defined translational and orientational order depending on the system parameters [15, 16]. The same morphological features observed in simulations are found in experimental samples of IPCs sedimented on a glass substrate [2]. In the bulk, results accumulated up to now have only scratched the surface of the many possibilities offered by IPCs to materials design. So far, the tendency towards two-dimensional ordering has been observed also in three-dimensions: different types of crystals consisting of parallel monolayers were found to be stable or even self-assembly in a wide region of the phase diagram [17–19]. More details on the assembly of these systems are reported in a recent review paper [20].

In this chapter, we thoroughly describe how to derive a coarse-grained description of the effective interactions between pairs of IPCs with an arbitrary surface pattern. The modeling was originally developed to describe IPCs carrying two polar patches with identical size and charge [1] and was subsequently extended to richer surface morphologies: in particular, it was thoroughly carried out for two patches with different charge and/or size [21]. The coarse-grained model of the pair potential between IPCs is derived by combining a mean field theoretical description based on

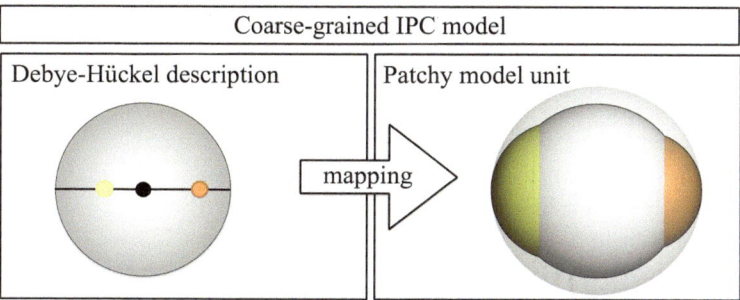

Fig. 2.1 The derivation of the coarse-grained model for the effective interaction between pairs of IPCs develops along two parallel paths: the Debye-Hückel description (left) and the patchy model (right); the resulting coarse-grained model is a patchy model whose parameters are chosen via a mapping from the former to the latter. The depicted IPC on the left is a dielectric sphere with a discrete and asymmetric charge distribution in its interior composed of three aligned and possibly different charges (yellow, black, and orange, from the left to the right). The depicted IPC on the right has two patches of different size and possibly different charge: the colloid is represented by the gray central sphere, while the two patches are represented by the yellow and orange spherical caps; the colloid interaction range is represented by the light gray halo. It is worth noting that IPCs are spherical units: the colored spherical caps represent the interaction range of the off-center sites located inside the colloidal surface

the Debye-Hückel theory with the design of a simple patchy model (see the left side of Fig. 2.1). The resulting coarse-grained model (1) has exactly the same symmetries as the underlying microscopic system, (2) maintains the mixed attractive/repulsive character of the former, depending on inter-particle orientations, and (3) allows for a simple computational approach.

The chapter is organized as follows: in Sect. 2.2 we derive the Debye-Hückel description, in Sect. 2.3 we introduce the simple patchy model, in Sect. 2.4 we connect the Debye-Hückel and the patchy approaches in order to derive our IPC coarse-grained model, i.e., a patchy model with parameters set by the features of the underlying microscopic system, in Sect. 2.5 we draw our concluding remarks.

2.2   The Debye-Hückel Description

We consider an IPC surrounded by a liquid solvent, such as water. According to Gauss' law, the electrostatic field outside a dielectric spherical colloid with a homogeneous surface charge is identical to the field generated by a point charge positioned in the center of the sphere. When a colloid has a heterogeneously charged surface, the charge of the different areas can be replaced by a discrete distribution of point charges positioned at the respective centers of charge inside the particle (see the left side of Fig. 2.1): the non-spherically symmetric distribution of discrete charges inside the colloidal particle must reproduce the symmetries of the surface pattern. It is worth noting that this approach is very general and is always reliable as long as the centers of charge of the different surface areas do not lie outside the colloidal particle. We apply the described approach to particles decorated by an arbitrary number of charged regions. In particular, we outline the general path for IPCs with n_p patches at well-defined positions.

In the following, we first derive the screened electrostatic potential generated by a single particle with a discrete distribution of charges (Sect. 2.2.1): the potential is calculated both inside (Sect. 2.2.1.1) and outside (Sect. 2.2.1.2) the colloid by expanding it in terms of spherical harmonics; then, by imposing electrostatic boundary conditions (Sect. 2.2.1.3), a set of linear equations for the expansion coefficients is obtained. By numerically solving the system of equations, the single particle potential is obtained. Based on this information, we determine the pair potential between two identical particles (Sect. 2.2.2): first the potential energy due to the presence of an IPC in the screened electrostatic field generated by another IPC is calculated, then the symmetric contribution is calculated, finally, the total interaction energy for a given particle–particle configuration is obtained as the average value over the two contributions.

For IPCs with two identical polar patches (Sect. 2.2.3), the system of differential equations leading to the single particle potential can be solved either analytically (relying on approximating the ratio between spherical Bessel functions of consecutive orders with a Yukawa-like term) or numerically. The advantage of the Yukawa-like approximation consists in providing an analytic expression for the pair potential that is immediately recognizable as a generalized Derjaguin-Landau-Verwey-Overbeek (DLVO) potential [22].

2.2.1 The Electrostatic Potential Around a Single IPC

2.2.1.1 Calculation of the Potential Inside the Colloid

We consider a reference system centered at the particle center and we make use of the spherical coordinates r, θ, and φ. The heterogeneously charged colloid is described as a dielectric sphere with a discrete charge distribution $\rho(r, \theta, \varphi)$ in its interior. The co- and counter-ions of the electrolyte solution cannot penetrate inside the particle. For the sake of simplicity, we set the elementary charge to unity and assume that the dielectric permittivity, ε, has the same value both inside and outside the colloid. Gaussian units are used in the following.

The potential $\Phi^{(1)}(r, \theta, \varphi)$ generated by the distribution of the charges inside the particle satisfies Poisson's equation

$$\Delta \Phi^{(1)}(r, \theta, \varphi) = -\frac{4\pi}{\varepsilon} \rho(r, \theta, \varphi), \qquad (2.1)$$

where the index (1) stands for "inside."

The general solution of Eq. (2.1) is the sum of the solution of the corresponding homogeneous equation (also known as Laplace's equation) plus a particular solution of the inhomogeneous equation, i.e.,

$$\Phi^{(1)}(r, \theta, \varphi) = \Phi^{(1)}_{\text{hom}}(r, \theta, \varphi) + \Phi^{(1)}_{\text{part}}(r, \theta, \varphi), \tag{2.2}$$

where $\Phi^{(1)}_{\text{hom}}(r, \theta, \varphi)$ can be written as

$$\Phi^{(1)}_{\text{hom}}(r, \theta, \varphi) = \sum_{\ell=0}^{\infty} \sum_{m=-\ell}^{+\ell} \left[A_{\ell m} r^{\ell} + B_{\ell m} r^{-\ell-1} \right] Y_{\ell m}(\theta, \varphi), \tag{2.3}$$

the $Y_{\ell m}(\theta, \varphi)$ being the spherical harmonics. Since the potential inside the colloidal sphere should not diverge at $r = 0$, the coefficients $B_{\ell m}$ are set to zero, i.e.,

$$\Phi^{(1)}_{\text{hom}}(r, \theta, \varphi) = \sum_{\ell=0}^{\infty} \sum_{m=-\ell}^{+\ell} A_{\ell m} r^{\ell} Y_{\ell m}(\theta, \varphi). \tag{2.4}$$

A particular solution of Eq. (2.1) depends on the specific charge distribution $\rho(r, \theta, \varphi)$. We sketch in the following two different examples.

- For IPCs with two, possibly different, polar patches and one equatorial region, the charge density can be seen as the result of a central charge, Z_c (the charge of the bare colloid) plus two out-of-center charges, Z_{p_1} and Z_{p_2} (the charges of the patches); the latter ones are located at distances a_1 and a_2 from the particle center, in directions opposite to each other. If the coordinate system is fixed such that one patch is located at $\varphi = 0$ and $\theta = \frac{\pi}{2}$ (and the other at $\varphi = \pi$ and $\theta = \frac{\pi}{2}$), the charge density in spherical coordinates reads as

$$\rho(r, \theta, \varphi) = Z_c \delta(\mathbf{r})$$
$$+ Z_{p_1} \frac{1}{a_1^2} \delta(r - a_1) \delta\left(\theta - \frac{\pi}{2} \right) \delta(\varphi)$$
$$+ Z_{p_2} \frac{1}{a_2^2} \delta(r - a_2) \delta\left(\theta - \frac{\pi}{2} \right) \delta(\varphi - \pi). \tag{2.5}$$

As a consequence, $\Phi^{(1)}_{\text{part}}(r, \theta, \varphi)$ is given by

$$\Phi^{(1)}_{\text{part}}(r, \theta, \varphi) = \frac{4\pi}{\varepsilon} \sum_{\ell=0}^{\infty} \sum_{m=-\ell}^{+\ell} \frac{1}{2\ell+1} \left[Z_{p_1} \frac{r_<^{\ell}}{r_>^{\ell+1}} Y_{\ell m}^*\left(\frac{\pi}{2}, 0 \right) \right.$$
$$\left. + Z_{p_2} \frac{r_<^{\ell}}{r_>^{\ell+1}} Y_{\ell m}^*\left(\frac{\pi}{2}, \pi \right) \right] Y_{\ell m}(\theta, \varphi) + \frac{4\pi}{\varepsilon} Z_c \frac{1}{r} Y_{00}^* Y_{00}, \tag{2.6}$$

where $r_> = {}^{\min}_{\max}\{a_\lambda, r\}$ with $\lambda = 1$ or 2 for the first and the second term in the square brackets, respectively, the star denotes complex conjugation and $Y_{00} = 1/\sqrt{4\pi}$.

- For IPCs with three, possibly different, patches, the charge density can be seen as the result of the central charge, Z_c, plus three out-of-center charges, Z_{p_1}, Z_{p_2}, and Z_{p_3}, located at (a_1, θ_1, ϕ_1), (a_2, θ_2, ϕ_2), and (a_3, θ_3, ϕ_3), respectively. The resulting charge density is thus

$$\rho(r, \theta, \varphi) = Z_c \delta(\mathbf{r})$$

$$+ Z_{p_1} \frac{1}{a_1^2} \delta(r - a_1) \delta(\theta - \theta_1) \delta(\varphi - \phi_1)$$

$$+ Z_{p_2} \frac{1}{a_2^2} \delta(r - a_2) \delta(\theta - \theta_2) \delta(\varphi - \phi_2)$$

$$+ Z_{p_3} \frac{1}{a_3^2} \delta(r - a_3) \delta(\theta - \theta_3) \delta(\varphi - \phi_3), \tag{2.7}$$

leading to the following particular solution of Eq. (2.1)

$$\Phi_{\text{part}}^{(1)}(r, \theta, \varphi) = \frac{4\pi}{\varepsilon} \sum_{\ell=0}^{\infty} \sum_{m=-\ell}^{+\ell} \frac{1}{2\ell + 1} \left[Z_{p_1} \frac{r_<^\ell}{r_>^{\ell+1}} Y_{\ell m}^* (\theta_1, \phi_1) \right.$$

$$\left. + Z_{p_2} \frac{r_<^\ell}{r_>^{\ell+1}} Y_{\ell m}^* (\theta_2, \phi_2) + Z_{p_3} \frac{r_<^\ell}{r_>^{\ell+1}} Y_{\ell m}^* (\theta_3, \phi_3) \right] Y_{\ell m}(\theta, \varphi)$$

$$+ \frac{4\pi}{\varepsilon} Z_c \frac{1}{r} Y_{00}^* Y_{00}. \tag{2.8}$$

where $r_> = {}^{\min}_{\max}\{a_\lambda, r\}$ with $\lambda = 1, 2$, or 3 for the first, the second, and the third term in the square brackets, respectively.

If all patches are located on the equatorial plane of the colloid and adjacent patches enclose equal angles, then $\theta_1 = \theta_2 = \theta_3 = \pi/2$ and $\phi_1 = 0, \phi_2 = 2\pi/3$ and $\phi_3 = 4\pi/3$ [21].

2.2.1.2 Calculation of the Potential Outside the Colloid

Outside the colloid the dielectric surrounding medium has to be taken into account: it consists of a large number of co- and counter-ions which—at equilibrium— follow the Boltzmann statistics. The non-linear differential equation describing the electrostatic potential in solution, $\Phi^{(2)}(r, \theta, \varphi)$, is known as the Poisson-Boltzmann equation [23] and reads as

$$\Delta \Phi^{(2)}(r, \theta, \varphi) = -\frac{4\pi}{\epsilon} \sum_{i=1}^{n} Z_i \rho_i \exp\left(-\frac{Z_i \Phi^{(2)}(r, \theta, \varphi)}{k_B T}\right), \tag{2.9}$$

where the index (2) stands for "outside," T is the temperature, k_B is the Boltzmann constant, i is the index running over all n species of ions in the solvent, ρ_i is the bulk density of species i, and Z_i is the valence of species i.

At low electrostatic potentials the Debye-Hückel approximation can be used. The Debye-Hückel approach consists in linearizing Eq. (2.9) by taking the Taylor expansion of the exponential truncated at the first order, i.e.,

$$\exp\left(-\frac{Z_i \Phi^{(2)}(r, \theta, \varphi)}{k_B T}\right) \approx 1 - \frac{Z_i \Phi^{(2)}(r, \theta, \varphi)}{k_B T}. \tag{2.10}$$

Because of the electroneutrality of the solution, the Poisson-Boltzmann equation for the potential outside the colloid becomes the so-called Helmholtz equation

$$\Delta \Phi^{(2)}(r, \theta, \varphi) = \kappa^2 \Phi^{(2)}(r, \theta, \varphi), \tag{2.11}$$

where κ is the inverse Debye screening length that encodes the characteristic features of the solvent; here

$$\kappa^2 = \frac{4\pi}{\epsilon k_B T} \sum_{i=1}^{n} \rho_i Z_i^2. \tag{2.12}$$

The Debye-Hückel approximation is strictly valid when $Z_i \Phi^{(2)}(r, \theta, \varphi) \ll k_B T$ (which means that in water at room temperature the potential at the surface must be lower than $\approx 26\,\text{meV}$). In order to check if the approximation is reasonable, one should check that the term $\left[Z_i \Phi^{(2)}(r, \theta, \varphi)/k_B T\right]$—where Z_i is the charge of the ionic species and $k_B T$ is calculated at room temperature—is small enough. To do so, once the potential $\Phi^{(2)}(r, \theta, \varphi)$ is known, it is important to remember that each charge must be multiplied by the unit of charge (while, for simplicity in the formulation, we have set it to unity throughout this chapter). By varying the different parameters that characterize in the potential (namely, the colloidal size, the effective charges of the different surface areas, the geometric parameters defining the charge distribution and the Debye screening length), it is possible to determine in which parameter windows the linear approximation is expected to be reliable. For example, given the charge distribution and fixed the effective charges, it is possible to determine the range of particle sizes for which the linear approximation is valid at different screening conditions [1].

In its most common form, the right-hand side of Eq. (2.11) has opposite sign, thus using the substitution $\kappa \to i\kappa$,[1] where i is the complex unit, in the solution of

---

[1] We note that, if we do not make the substitution $\kappa \to i\kappa$, we have to solve the spherical Bessel differential equation with a negative separation constant. In this case, $\Phi^{(2)}(r, \theta, \varphi)$ can be expressed in terms of modified spherical Bessel functions of the first and second kind, i.e.,

$$\Phi^{(2)}(r, \theta, \varphi) = \sum_{\ell=0}^{\infty} \sum_{m=-\ell}^{+\ell} \left[C'_{\ell m} i_\ell(\kappa r) + D'_{\ell m} k_\ell(\kappa r)\right] Y_{\ell m}(\theta, \varphi). \tag{2.13}$$

the standard Helmholtz equation, we arrive at [24]

$$\Phi^{(2)}(r, \theta, \varphi) = \sum_{\ell=0}^{\infty} \sum_{m=-\ell}^{+\ell} [C_{\ell m} j_\ell(i\kappa r) + D_{\ell m} y_\ell(i\kappa r)] Y_{\ell m}(\theta, \varphi), \qquad (2.15)$$

where $j_\ell(i\kappa r)$ and $y_\ell(i\kappa r)$ are the spherical Bessel functions of the first and second kind, respectively, defined as [24]

$$j_\ell(i\kappa r) = (i)^\ell (\kappa r)^\ell \left(\frac{1}{\kappa r} \frac{d}{d(\kappa r)} \right)^\ell \frac{\sinh \kappa r}{\kappa r} \qquad (2.16)$$

$$y_\ell(i\kappa r) = -(i)^\ell (\kappa r)^\ell \left(\frac{1}{\kappa r} \frac{d}{d(\kappa r)} \right)^\ell \frac{\cosh \kappa r}{\kappa r} \qquad (2.17)$$

with $\ell = 0, 1, 2, 3, \ldots$. Since $\Phi^{(2)}(r, \theta, \varphi)$ has to vanish for $r \to \infty$, while $j_\ell(i\kappa r)$ and $y_\ell(i\kappa r)$ both diverge at large distances, we introduce a suitable linear combination of the two sets of functions, namely the spherical Bessel functions of the third kind (also referred to as spherical Hankel functions of the first kind), defined as [24]

$$h_\ell^{(1)}(i\kappa r) = j_\ell(i\kappa r) + i y_\ell(i\kappa r) = -(i)^\ell (\kappa r)^\ell \left(\frac{1}{\kappa r} \frac{d}{d(\kappa r)} \right)^\ell \frac{e^{-\kappa r}}{\kappa r}, \qquad (2.18)$$

so that the potential outside the colloid can be written as

$$\Phi^{(2)}(r, \theta, \varphi) = \sum_{\ell=0}^{\infty} \sum_{m=-\ell}^{+\ell} E_{\ell m} h_\ell^{(1)}(i\kappa r) Y_{\ell m}(\theta, \varphi), \qquad (2.19)$$

or equivalently (see footnote 1) as

$$\Phi^{(2)}(r, \theta, \varphi) = \sum_{\ell=0}^{\infty} \sum_{m=-\ell}^{+\ell} E'_{\ell m} \frac{K_{\ell+1/2}(\kappa r)}{\sqrt{r}} Y_{\ell m}(\theta, \varphi), \qquad (2.20)$$

where $K_{\ell+1/2}(\kappa r)$ are modified Bessel functions of the second kind.

Remembering that $i_\ell(\kappa r) = \sqrt{\frac{\pi}{2\kappa r}} I_{\ell+1/2}(\kappa r)$ and $k_\ell(\kappa r) = \sqrt{\frac{\pi}{2\kappa r}} K_{\ell+1/2}(\kappa r)$ [24], we obtain the same expression used in [25], i.e.,

$$\Phi^{(2)}(r, \theta, \varphi) = \sum_{\ell=0}^{\infty} \sum_{m=-\ell}^{+\ell} \left[C''_{\ell m} \frac{I_{\ell+1/2}(\kappa r)}{\sqrt{r}} + D''_{\ell m} \frac{K_{\ell+1/2}(\kappa r)}{\sqrt{r}} \right] Y_{\ell m}(\theta, \varphi). \qquad (2.14)$$

Note that, since the potential must vanish at infinity, $C''_{\ell m}$ must be zero.

2.2.1.3 Linking the Two Potentials via Boundary Conditions

To obtain the full potential over the whole space, we must solve the system of differential equations (and evaluate the $A_{\ell m}$ and $E_{\ell m}$ coefficients); thus, we need to consider the proper boundary conditions.

First, the potential must be continuous at the particle surface (i.e., at $r = \sigma_c$)

$$\Phi^{(1)}(r, \theta, \varphi)|_{r=\sigma_c} \equiv \Phi^{(2)}(r, \theta, \varphi)|_{r=\sigma_c}. \tag{2.21}$$

Second, the tangential component of the electrostatic field has to be continuous. Since the partial derivatives with respect to the polar and azimuthal coordinates act only on the spherical harmonics,[2] they produce sets of linearly independent functions that lead to the same relations for the coefficients; thus, only the θ derivative of the potential is needed to express the required boundary condition, that is

$$\partial_\theta \Phi^{(1)}(r, \theta, \varphi)|_{r=\sigma_c} \equiv \partial_\theta \Phi^{(2)}(r, \theta, \varphi)|_{r=\sigma_c}. \tag{2.23}$$

Finally, in the absence of surface charges, the normal component of the displacement field must be continuous. Since the relative permittivity ε is taken to be the same inside and outside of the colloid, this boundary condition can be written as

$$\partial_r \Phi^{(1)}(r, \theta, \varphi)|_{r=\sigma_c} \equiv \partial_r \Phi^{(2)}(r, \theta, \varphi)|_{r=\sigma_c}. \tag{2.24}$$

When the dielectric discontinuity between the colloid and the surrounding medium is taken into account, then on the right/left side of Eq. (2.24) the dielectric constant of the colloid/solvent must be included [25].

We now insert the expansions of $\Phi^{(1)}(r, \theta, \varphi)$—see Eq. (2.2)—and of $\Phi^{(2)}(r, \theta, \varphi)$—see Eq. (2.19)—into these relations. Since the spherical harmonics are linearly independent, we obtain for each index combination (ℓ, m) an equation for the corresponding, yet unknown coefficient; the series expansions have to be truncated at a suitable upper limit, ℓ_{max}, for practical reasons. Once these coefficients are known, the resulting electrostatic potential around a single IPC is thus available. Being interested only in the potential outside the colloid, we will omit in the following the superscript 2 and refer to the screened electrostatic potential in the region of interest as $\Phi(r, \theta, \varphi)$.

2.2.2 The Effective Potential Between Two IPCs

Once the single particle potential $\Phi(r, \theta, \varphi)$ is known, the effective interaction energy between two identical IPCs can be calculated according to the following

---

[2]We remind that the Laplacian operator in spherical coordinates is

$$\frac{1}{r^2}\frac{\partial}{\partial r}\left(r^2\frac{\partial}{\partial r}\right) + \frac{1}{r^2 \sin^2\varphi}\frac{\partial^2}{\partial\theta^2} + \frac{1}{r^2 \sin\varphi}\frac{\partial}{\partial\varphi}\left(\sin\varphi\frac{\partial}{\partial\varphi}\right). \tag{2.22}$$

procedure. We consider two colloids, with indices i and j, separated by a distance r_{ij} and with mutual orientation Ω_{ij}. Since the effective pair potential, $\psi(r_{ij}, \Omega_{ij})$, has to be symmetric with respect to the indices i and j, we calculate both (a) the potential energy of particle j in the electrostatic field generated by particle i, labelled as ψ_{ij}, and (b) the potential energy of particle i in the electrostatic field generated by particle j, labelled as ψ_{ji}; the total interaction energy for a given particle–particle configuration is defined as the average value over the contribution (a) and (b). In order to calculate (a), particle i is considered as the point-like source of the field, while particle j is considered as a dielectric sphere with a distribution of point charges inside; vice versa for the calculation of (b). This means that the distances and the angles to be considered are those between the center of the source particle and the discrete charges inside the probe particle (see Fig. 2.2). The resulting pair potential is given by the sum of the screened Coulomb potentials generated by the source sphere and centered at the positions of the discrete charges inside the probe sphere; in the sum, each contribution is multiplied by the magnitude of the corresponding charge at that point. Namely,

$$\psi(r_{ij}, \Omega_{ij}) = \frac{1}{2}(\psi_{ij} + \psi_{ji}) \tag{2.25}$$

$$= \frac{1}{2}\left[\sum_{\lambda=0}^{n_p} Z_\lambda^j \frac{\exp(\kappa\sigma_c)}{1+\kappa\sigma_c}\Phi(r_{ij}^\lambda, \theta_i^\lambda, \varphi_i^\lambda) + \sum_{\lambda=0}^{n_p} Z_\lambda^i \frac{\exp(\kappa\sigma_c)}{1+\kappa\sigma_c}\Phi(r_{ji}^\lambda, \theta_j^\lambda, \varphi_j^\lambda)\right],$$

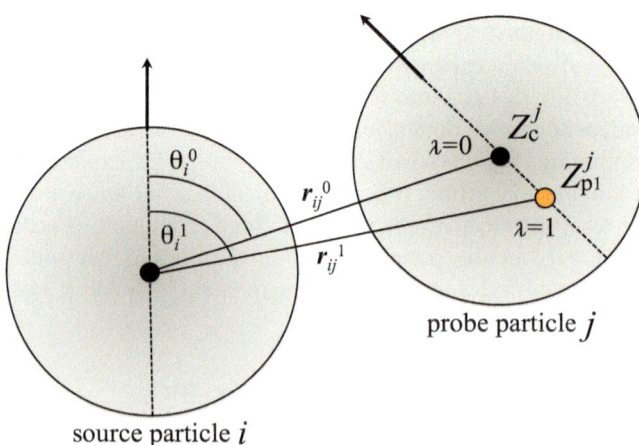

Fig. 2.2 To calculate the potential energy of the probe particle j in the electrostatic field generated by the source particle i, particle j is considered as a sphere carrying a discrete charge distribution in its interior (here, a central charge, Z_c^j, and one out-of-center charge, Z_{p1}^j), while particle i is considered as a point-like source. For the sake of representation, the reference system is centered at the position of particle i and the orientational unit vectors of the two particles are in the same plane, thus $\varphi_i^\lambda = 0$ for both λ-values, while θ_i^λ and r_{ij}^λ are as represented in the figure

where λ runs over the central charge (corresponding to the bare colloid) plus the n_p out-of-center charges (corresponding to the patches): $\lambda = 0$ identifies the central charge (i.e., $Z_0^i \equiv Z_c^i$), while $\lambda > 0$ identifies the out-of-center charges (i.e., $Z_1^i \equiv Z_{p1}^i$ and so on); for the definition of distances and angles, we refer to Fig. 2.2. We note that, similar to the derivation of the DLVO potential, we take into account the fact that microscopic co- and counter-ions cannot penetrate the IPC by replacing the bare charges Z_λ^i with effective charges $Z_\lambda^i \frac{\exp(\kappa\sigma_c)}{1+\kappa\sigma_c}$.

2.2.3 Case Study: Two Identical Patches

For IPCs with two identical polar patches, i.e., $Z_{p1} = Z_{p2} \equiv Z_p$ and $a_1 = a_2 \equiv a$, it is possible to take advantage of the cylindrical symmetry of the discrete charge distribution [1].

To calculate the single particle potential, we assume that the three centers of charge within the IPC lie on the z axis and put the origin of the reference system in the center of the colloidal particle. Due to the azimuthal symmetry of the charge distribution, the charge density depends only on r and θ and thus, Eq. (2.5) becomes

$$\rho(r, \theta) = Z_c \delta(\mathbf{r}) + Z_p \left[\delta(\mathbf{r} - a\hat{z}) + \delta(\mathbf{r} + a\hat{z}) \right]. \tag{2.26}$$

The resulting electrostatic expansions for the single particle potential can thus be expressed in terms of Legendre polynomials $P_\ell(\cos\theta)$. In particular, the general solution of Eq. (2.1)—i.e., expression (2.2) for the potential inside the colloid—becomes[3]

$$\Phi^{(1)}(r, \theta) = \sum_{\ell=0}^{\infty} A_\ell r^\ell P_\ell(\cos\theta) + \frac{Z_c}{\epsilon r} + \frac{2Z_p}{\epsilon} \sum_{\ell=0}^{\infty}{}' \frac{r_<^\ell}{r_>^{\ell+1}} P_\ell(\cos\theta), \tag{2.28}$$

where the prime indicates that the sum runs only over even index values ℓ and $r_<$ ($r_>$) is the larger (smaller) between r and a; in contrast, the general solution of Eq. (2.11)—i.e., expression (2.20) for the potential outside the colloid—becomes

$$\Phi^{(2)}(r, \theta) = \sum_{\ell=0}^{\infty} E_\ell \frac{K_{\ell+1/2}(\kappa r)}{\sqrt{r}} P_\ell(\cos\theta), \tag{2.29}$$

where $K_{\ell+1/2}(\cos\theta)$ are modified spherical Bessel functions of the third kind [24].

---

[3]Summations over the azimuthal index m are carried out explicitly by making use of the addition theorem of spherical harmonics:

$$\sum_{m=-\ell}^{\ell} Y_{\ell m}^*(\theta', \varphi') Y_{\ell m}(\theta, \varphi) = \frac{2\ell+1}{4\pi} P_\ell(\mathbf{n}' \cdot \mathbf{n}) \tag{2.27}$$

where the star denotes complex conjugation and $\mathbf{n}'$ and $\mathbf{n}$ are unit vectors in the (θ', φ') and (θ, φ) directions, respectively (i.e., the source and observation point directions).

By using the boundary conditions specified in Sect. 2.2.1.3, the system of differential equations can be analytically solved, resulting in the following potential for the region outside the colloid

$$\Phi(r,\theta) = \frac{Z_c + 2Z_p}{\epsilon} \frac{\exp(\kappa\sigma_c)}{1 + \kappa\sigma_c} \frac{\exp(-\kappa r)}{r}$$

$$+ \frac{2Z_p}{\epsilon} \sum_{\ell=2}^{\infty}{}' \left(\frac{a}{\sigma_c}\right)^{\ell} \frac{(2\ell+1)}{\kappa\sigma_c\sqrt{r\sigma_c}} \frac{K_{\ell+1/2}(\kappa r)}{K_{\ell+3/2}(\kappa\sigma_c)} P_\ell(\cos\theta). \quad (2.30)$$

As Eq. (2.30) involves both Legendre polynomials and modified spherical Bessel functions of the third kind, the analytic calculation of the effective pair interaction between two IPCs can be complex. In order to proceed with the analytical approach, we can approximate the ratio between Bessel functions of consecutive orders by its asymptotic term, i.e.,[4]

$$\frac{1}{\kappa\sigma_c\sqrt{r\sigma_c}} \frac{K_{\ell+1/2}(\kappa r)}{K_{\ell+3/2}(\kappa\sigma)} \approx \frac{\exp(\kappa\sigma_c)}{1 + \kappa\sigma_c} \frac{\exp(-\kappa r)}{r}. \quad (2.32)$$

We observe that this approximation is reasonable under high screening conditions, i.e., for $\kappa\sigma_c \gg 1$ [1]. Thanks to this approximation, $\Phi(r,\theta)$ can be factorized in a radially symmetric Yukawa contribution and an angle dependent factor that takes into account the non-spherically symmetric charge distribution

$$\Phi(r,\theta) = \left[Z_c + 2Z_p \sum_{\ell=0}^{\infty}{}' \left(\frac{a}{\sigma_c}\right)^{\ell} (2\ell+1)P_\ell(\cos\theta)\right] \frac{\exp(\kappa\sigma_c)}{1 + \kappa\sigma_c} \frac{\exp(-\kappa r)}{\epsilon r}. \quad (2.33)$$

We note that, the approximated electrostatic potential of Eq. (2.33) is found to be reasonably close to the analytic potential of Eq. (2.30), even at short distances, i.e., on the particle surface [1]. Moreover, the approximated electrostatic potential of Eq. (2.33) has also been compared to the numerical solution, proving itself to be a very reasonable approximation also at small $\kappa\sigma$ values [21]. Such a simplification allows us to analytically derive the effective pair potential between two IPCs.

We now derive the effective interaction between two IPCs, with indices i and j, according to Eq. (2.25). The first term appearing in Eq. (2.25), ψ_{ij}, can be written in a DLVO-like expression, i.e.,

$$\psi_{ij} = \frac{Q(r_{ij}, \Omega_{ij})}{\epsilon} \left[\frac{\exp(\kappa\sigma_c)}{1 + \kappa\sigma_c}\right]^2 \frac{\exp(-\kappa r_{ij})}{r_{ij}}, \quad (2.34)$$

---

[4]We use that [24]:

$$\sqrt{\frac{\pi}{2}} \frac{K_{\ell+1/2}(\kappa r)}{\kappa r} \approx \frac{\pi}{2} \frac{\exp(-\kappa r)}{\kappa r}. \quad (2.31)$$

where $Q(r_{ij}, \Omega_{ij})$ is a spatially and orientationally dependent factor, which takes into account all the charge valences involved in the interaction in the following way

$$Q(r_{ij}, \Omega_{ij}) = \left[Z_c^2 + 2Z_cZ_p \sum_{\ell=0}^{\infty}{}' (2\ell + 1) \left(\frac{a}{\sigma_c} \right)^{\ell} P_{\ell}(\cos\theta_i^0) \right] \qquad (2.35)$$

$$+ \left[Z_cZ_p + 2Z_p^2 \sum_{\ell=0}^{\infty}{}' (2\ell + 1) \left(\frac{a}{\sigma_c} \right)^{\ell} P_{\ell}(\cos\theta_i^1) \right] \frac{\exp\left[-\kappa r_{ij}(\xi_1 - 1) \right]}{\xi_1}$$

$$+ \left[Z_cZ_p + 2Z_p^2 \sum_{\ell=0}^{\infty}{}' (2\ell + 1) \left(\frac{a}{\sigma_c} \right)^{\ell} P_{\ell}(\cos\theta_i^2) \right] \frac{\exp\left[-\kappa r_{ij}(\xi_2 - 1) \right]}{\xi_2},$$

where the prime denotes a sum over even power of ℓ only. In the above expression, we have used the dimensionless quantities $\xi_\lambda = r_{ij}^\lambda / r_{ij}$ with $\lambda = 1, 2$ (since $\xi_0 = 1$). In the same way, we can write an expression for the second term appearing in Eq. (2.25), ψ_{ji}, thus obtaining the total symmetric pair potential between two IPCs, $\psi(r_{ij}, \Omega_{ij})$.

It is worth noting that the following DLVO limits can easily be recovered [1]. Trivially, if $Z_p \to 0$, then $Q \to Z_c^2$ and Eq. (2.34) reduces to the DLVO interaction between two colloids, each of them carrying a homogeneously distributed charge Z_c. Furthermore, if $Z_p \neq 0$ and $a \to 0$, then the exponential factors in Q tend to unity as $\xi_{1,2} \to 1$, and the only terms to survive in Eq. (2.35) are those for $\ell = 0$; thus Eq. (2.34) reduces again to a DLVO-like potential between two colloids, each of them carrying a homogeneously distributed charge $(Z_c + 2Z_p)$.

2.3   The Patchy Model

The patchy model is designed to reproduce the same symmetries as the underlying microscopic system (see the right side of Fig. 2.1) and is characterized by three independent sets of parameters: two sets of *geometric* parameters, i.e., the interaction ranges of the different surface regions and their surface extents, and one set of *energy* parameters, i.e., the characteristic interaction strengths between the different surface areas.

The model features a hard spherical particle of radius σ_c carrying n_p interaction sites, each placed at distance a_λ from the particle center, with $\lambda = 1, \ldots, n_p$; such a distance is always smaller than σ_c so that the sites are always located inside the colloid. As a consequence, the corresponding site interaction sphere, with radius $\sigma_{p\lambda}$, extends only partially outside the hard core particle, defining in this way the patch λ; the respective surface extension of the patch is characterized by the half opening angle γ_λ (see Fig. 2.3) .

The first set of geometric parameters (i.e., the interaction ranges between the different surface areas) is chosen in a straightforward manner: since the characteristic interaction distances in the microscopic system are determined by the electrostatic

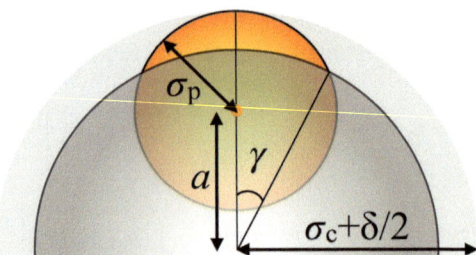

Fig. 2.3 The model geometric parameters are the size of the colloids σ_c, the particle interaction range δ, and the patch opening angle γ (as labeled); the latter two are determined once the position of the interaction sites a and the site interaction range σ_p (as labeled) are chosen. The particle core is represented in dark gray, the patch interaction sphere in orange, and the particle interaction sphere in light grey

screening of the surrounding solvent, all the surface regions of the colloid are assumed to have the same interaction range, δ. The interaction sphere of the bare colloid has thus radius $\sigma_c + \delta/2$. As a consequence, the following relations hold by construction for each patch λ (see Fig. 2.3)

$$\frac{\delta}{2} = a_\lambda + \sigma_{p\lambda} - \sigma_c \tag{2.36}$$

$$\cos \gamma_\lambda = \frac{\sigma_c^2 + a_\lambda^2 - \sigma_{p\lambda}^2}{2\sigma_c a_\lambda}. \tag{2.37}$$

The second set of geometric parameters to be fixed is the set of patch sizes, which can be freely chosen to reproduce the surface features of experimental IPCs [2]. By virtue of the aforementioned constraints, once a_λ and $\sigma_{p\lambda}$ are defined, also δ and γ_λ are fixed, so that the model is characterized only by the physical parameters δ and γ_λ, while σ_c is the unit of length.

Finally, the third set of parameters defining our patchy model is the set of values for the energy strengths between the differently charged surface regions. Once the geometric parameters are set, the energy parameters are determined only by the charges associated to the different surface areas: while the center of the particle carries a charge Z_c, the sites carry a charge $Z_{p\lambda}$ each. These charges are responsible for the ratio between the attractive and repulsive contributions to the pair energy.

The specific form of the interparticle potential is based on the postulate that the different contributions to the pair interaction can all be factorized into a characteristic energy strength and a geometric weight factor, the latter one taking into account the distance and relative orientation between the interacting particles. More specifically, beyond the hard core repulsion, the pair potential between two IPCs at distance r_{ij} with a mutual orientation Ω_{ij} is given by

$$U(r_{ij}, \Omega_{ij}) = \begin{cases} \sum_{\alpha\beta} u_{\alpha\beta}^{ij} w_{\alpha\beta}^{ij}(r_{ij}, \Omega_{ij}) & \text{if} \quad 2\sigma_c < r_{ij} < 2\sigma_c + \delta \\ 0 & \text{if} \quad r_{ij} \geq 2\sigma_c + \delta \end{cases} \tag{2.38}$$

where i and j specify the particles, α and β specify either a patch (P) or the bare colloid (C) of the first and second IPC, respectively, while $w_{\alpha\beta}^{ij}$ and $u_{\alpha\beta}^{ij}$ are the geometric weight factor and the characteristic energy strength of the $\alpha\beta$ interaction, respectively. We note that, while the $u_{\alpha\beta}^{ij}$ are constant values, the $w_{\alpha\beta}^{ij}$—as well as the potential U—depend on both the inter-particle distance and the relative orientation of the two IPCs. For the sake of simplicity we omit in the following the explicit dependence of the $w_{\alpha\beta}^{ij}$ on r_{ij} and Ω_{ij} and we show in three steps how each $w_{\alpha\beta}^{ij}$ is calculated for all possible α and β combinations.

First of all, each $w_{\alpha\beta}^{ij}$ is proportional to the total overlap volume, $W_{\alpha\beta}^{ij}$, between all the interaction spheres of type α on particle i and all the interaction spheres of type β on particle j, i.e.,

$$w_{\alpha\beta}^{ij} = W_{\alpha\beta}^{ij}/W_{\text{Ref}}, \tag{2.39}$$

where W_{Ref} is the reference volume of the colloidal hard sphere, i.e., $W_{\text{Ref}} = \frac{4}{3}\pi\sigma_{\text{c}}^3$.

As a second step, one should consider that, since α and β can be either C or P, the total overlap volumes that must be calculated for two given particles i and j are, in general (i.e., without making any explicit distinctions between identical or different patches), of three types: W_{CC}^{ij}, W_{CP}^{ij}, and W_{PP}^{ij}. The most general expressions for these three quantities can be written as

$$W_{\text{CC}}^{ij} = W_{\text{C}_i\text{C}_j} \tag{2.40}$$

$$W_{\text{CP}}^{ij} = \sum_{\lambda=1}^{n_{\text{p}}}(W_{\text{C}_i\text{P}_j^\lambda} + W_{\text{P}_i^\lambda\text{C}_j}) \tag{2.41}$$

$$W_{\text{PP}}^{ij} = \sum_{\lambda,\zeta=1}^{n_{\text{p}}} W_{\text{P}_i^\lambda\text{P}_j^\zeta} \tag{2.42}$$

where λ and ζ denote the patch indices on particle i or j, respectively.

Each of the contributions appearing on the right side of Eqs. (2.40)–(2.42) is an overlap volume between two spheres, W_{12} (where 1 and 2 are either C or P), of radii R_1 and R_2, respectively, separated by a distance r_{12}. The explicit form of W_{12} can be written as

$$W_{12} = \begin{cases} 0 & \text{if } r_{\max} \leq r_{12} \\ \frac{\pi}{3}\left[\left(2R_1 + \frac{R_1^2-R_2^2+r_{12}^2}{2r_{12}}\right)\left(R_1 - \frac{R_1^2-R_2^2+r_{12}^2}{2r_{12}}\right)^2\right] + \\ \frac{\pi}{3}\left[\left(2R_2 - \frac{R_1^2-R_2^2-r_{12}^2}{2r_{12}}\right)\left(R_2 + \frac{R_1^2-R_2^2-r_{12}^2}{2r_{12}}\right)^2\right] & \text{if } r_{\min} \leq r_{12} \leq r_{\max} \\ \frac{4}{3}\pi R_<^3 & \text{if } r_{12} \leq r_{\min} \end{cases} \tag{2.43}$$

where $R_< = \min(R_1, R_2)$, $r_{\max} \equiv R_1 + R_2$ is the distance above which the two spheres do not overlap anymore, and $r_{\min} \equiv |R_1 - R_2|$ is the distance below which the two spheres completely overlap. In our particular case, r_{min} plays a role only for the CP interaction: due to the hard core repulsion, spheres of the same size are prevented from completely overlap, while a small sphere may happen to be totally included inside a big sphere. It is important to bear in mind that the distance r_{12} does not necessarily coincide with the distance between the two IPCs. Indeed, $r_{12} = r_{ij}$ only when $12 = CC$, while for $12 = CP$ and $= PP$, r_{12} is in general a function of r_{ij} and such a function depends on the relative orientation of the two IPCs. Consequently, the resulting $w_{\alpha\beta}^{ij}$ depend on both r_{ij} and Ω_{ij}.

We thus have obtained an expression for the effective interaction between two IPCs, that can be evaluated in a fast and efficient way and that can readily be used in numerical approaches: for two interacting IPCs the interaction energy is straightforwardly calculated once all the distances between the different interaction spheres are determined. The evaluation of the pair interaction between two IPCs does not require additional information.

2.4   The Coarse-Grained Description

While the patchy model described in Sect. 2.3 is still a toy model, the availability of the Debye-Hückel description reported in Sect. 2.2 allows us to choose the model parameters such that they are related to the physical properties of the underlying microscopic system, as described in the following.

As stated before, the model parameters that enter into the particle–particle interaction energy between pairs of IPCs are two sets of geometric parameters and one set of energy parameters. More specifically, the model pair interaction between IPCs is given by Eq. (2.38), where we need to evaluate the weight functions $w_{\alpha\beta}^{ij}(r_{ij}, \Omega_{ij})$ and the energy strengths $u_{\alpha\beta}^{ij}$: the weight functions for a pair of particles in a given configuration and with a given patch arrangement depend on the geometric parameters of the model (i.e., on δ and γ_λ), while the energy strengths are constants to be defined. We note that, for a given set of geometric parameters, the $u_{\alpha\beta}^{ij}$ are essentially related to the overall particle charge $Z_{tot} = Z_c + \sum_{\lambda=1}^{n_p} Z_\lambda$.

For what concerns the length scales, we assume that the model interaction range δ is proportional to the Debye screening length according to the following relation

$$\kappa\delta = n, \qquad (2.44)$$

where κ is determined by the screening conditions $\kappa\sigma_c = m$; thus, $\delta = \frac{n}{m}\sigma_c$, where m and n are not necessarily integer numbers. We note that usually the Debye screening length sets the characteristic interaction range of the DLVO potential (i.e., $\kappa\delta = 1$): here we just extend the relation between δ and κ to allow for a more quantitative evaluation of the characteristic interaction distances.

For what concerns the surface patterns, once δ is fixed, the size of a patch is defined either by the position of the patch center of charge or by the size of the corresponding interaction sphere. In other words, for a given δ, the angular patch

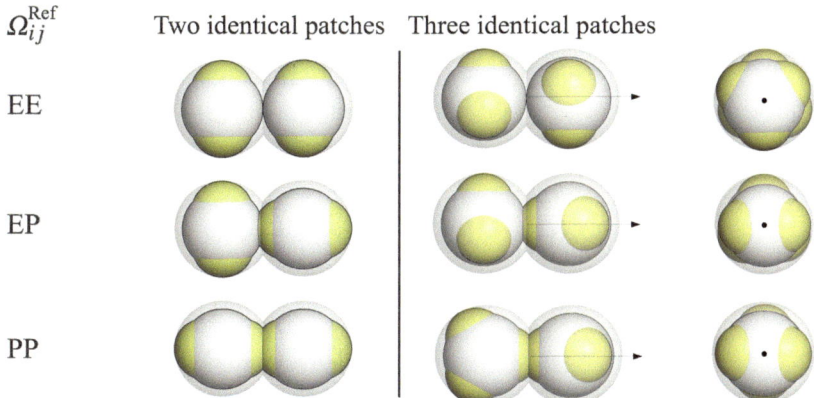

Fig. 2.4 Pairs of IPCs at contact in the three orientational configurations Ω_{ij}^{Ref} used as reference in the mapping between the Debye-Hückel potential and the patchy model potential; from the top to the bottom (as labeled on the left): the equatorial-equatorial (EE), the equatorial-polar (EP), and the polar-polar (PP) configuration. First column: side views of IPCs with two identical polar patches. Second/third column: side/top views of IPCs with three identical patches. Note that, for IPCs with three identical patches, the second particle can rotate with respect to the first one—preserving pure EE, EP, and PP interactions—around an axis depicted as a black arrow/dot in the side/top view

extents γ_λ are defined by the choice of either a_λ or $\sigma_{p\lambda}$ by virtue of Eqs. (2.36) and (2.37).

For what concerns the determination of the $u_{\alpha\beta}^{ij}$, we must take into account all the required interaction types: the general procedure is described in detail in the following for an arbitrary surface decoration and then some examples for specific surface patterns are proposed.

For each $\alpha\beta$ interaction, we first choose the most significant configuration of two IPCs such that the selected $\alpha\beta$ interaction is pure or at least predominant over the others; subsequently, for the chosen set of r_{ij} and Ω_{ij}, we perform a mapping between the Debye-Hückel effective potential—fully determined by the physical properties of the underlying microscopic system—and the patchy pair potential—characterized by the yet undetermined $u_{\alpha\beta}^{ij}$. The number of reference configurations to be considered depends on the number of $\alpha\beta$ interactions to be evaluated, which in turn depends on the number of differently charged or sized surface areas (see some examples in Fig. 2.4).

In [1] two mapping schemes have been proposed to fix the energy strengths. The first mapping scheme, named "tot," consists in matching the integral of the two potentials over their whole interaction range for two IPCs with a fixed mutual orientation (i.e., $\Omega_{ij} = \Omega_{ij}^{\text{Ref}}$), namely (omitting the explicit dependence of the potentials on the distance and mutual orientation between the two particles)

$$\frac{1}{\delta} \int_{2\sigma_c}^{\infty} \psi_{ij}\big|^{\Omega_{ij}=\Omega_{ij}^{\text{Ref}}} \mathrm{d}r_{ij} \equiv \frac{1}{\delta} \int_{2\sigma_c}^{\infty} U_{ij}\big|^{\Omega_{ij}=\Omega_{ij}^{\text{Ref}}} \mathrm{d}r_{ij}, \qquad (2.45)$$

where $\psi_{ij} = \psi(r_{ij}, \Omega_{ij})$ and $U_{ij} = U(r_{ij}, \Omega_{ij})$. The second mapping scheme, named "max", consists in matching the value of the two potentials for two IPCs at contact (i.e., $r_{ij} = 2\sigma_c$) with a fixed mutual orientation (i.e., $\Omega_{ij} = \Omega_{ij}^{\mathrm{Ref}}$), namely (omitting again the explicit dependence of the potentials on r_{ij} and Ω_{ij})

$$\psi_{ij}\Big|_{r_{ij}=2\sigma_c}^{\Omega_{ij}=\Omega_{ij}^{\mathrm{Ref}}} \equiv U_{ij}\Big|_{r_{ij}=2\sigma_c}^{\Omega_{ij}=\Omega_{ij}^{\mathrm{Ref}}} . \tag{2.46}$$

The two mapping schemes yield similar results; for this reason, in the literature, only the latter has been used (see, e.g., [15–19]).

As noted before, the evaluation of the $u_{\alpha\beta}^{ij}$ depends on the particle surface pattern, since the patch decoration determines both the number of $\alpha\beta$ interaction types to be considered and the choice of significant configurations to be used as reference. In general, when patches are highly symmetric, highly symmetric reference configurations are enough. For IPCs with two symmetric polar patches the number of $\alpha\beta$ interaction types is three (patch/patch, patch/non-patch, and non-patch/non-patch) and thus the three configurations depicted in the first column of Fig. 2.4, namely the polar–polar (PP), the equatorial–polar (EP), and the equatorial–equatorial (EE) configuration, are enough to fully describe the inter-particle potential [1]. For IPCs with three identical patches, then the number of $\alpha\beta$ interaction types is still three but the choice of the reference configurations is to some extent arbitrary. In [21] we suggested that, even if the most characteristic configurations needed to evaluate all the pure $\alpha\beta$ interactions are the ones depicted in the second and third column of Fig. 2.4 (labeled again EE, EP, and PP, respectively), it is equally valid to consider rotations of one of the two IPCs that preserve the pure overlap between the selected interaction spheres (as shown in Fig. 2.4); in the latter case, the average value of the Debye-Hückel contact energies of all equivalent configurations is used as the reference for the mapping to the patchy model contact energies. For the full description of IPCs with two asymmetric patches (either in size or in charge), six characteristic particle configurations are needed: in particular, in the PP and EP configurations the existence of two different patches must be taken into account [21]. For the full description of IPCs with three asymmetric patches, ten reference configurations with the pure $\alpha\beta$ interactions must be considered: one EE configuration, three EP configurations, and six PP configurations. Due to the reduced symmetry as compared to the identical patch case, averaging over all possible rotations of one IPC with respect to the other is advisable (the same rotation axis shown in Fig. 2.4 can be used).

It is worth noting that, once the coarse-grained potential is obtained, we normalize it in order to set a scale for the temperature: the unit of temperature is thus $e_{\min}/k_B$. The advantage of using reduced units is that many combinations of temperature, particle size, and minimum energy correspond to the same state point in reduced units [26]. In the case of IPCs with two or three identical patches, the pair interaction energy is normalized by the value corresponding to the energy of the EP configuration, i.e., $e_{\min} \equiv e_{\mathrm{PP}}$, while in the case of two different patches, the interaction energies are normalized by the most negative value occurring (i.e., the most attractive interaction), corresponding either to the EP_1 or the EP_2 configuration.

As an example, we consider now overall neutral IPCs with two identical polar patches and show the numerical, analytical, and coarse-grained potentials in Fig. 2.5. In particular, we consider a symmetric and in-line discrete charge distribution characterized by a central site, carrying a charge $Z_c = -180$, and two out-of-center sites positioned at $a_1 \equiv a_2 \equiv a = 0.22$ and carrying the same charge

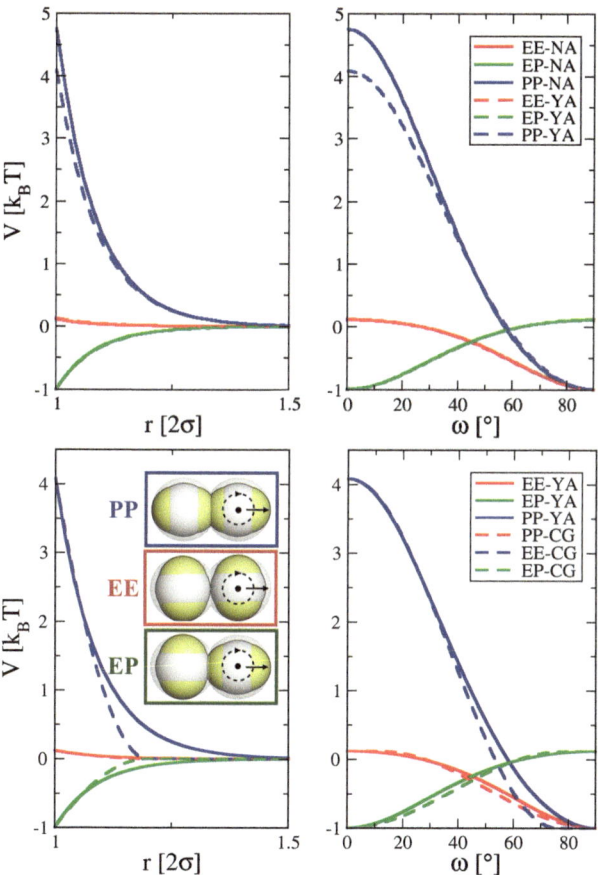

Fig. 2.5 Pair potentials between IPCs with two identical patches (parameters given in the text). The left/right panels show the radial/angular dependence of the potentials; the red, blue, and green continuous/dashed lines correspond to different initial configurations (shown as insets in the bottom-left panel): colloids at contact with PP, EE, and EP reciprocal orientation, respectively (as labelled). In all particle–particle configurations, the continuous black arrow indicates the translation direction of one IPC with respect to the other, while the dashed circle with one central dot represents the rotation of one IPC around an axis perpendicular to the plane. Top: comparison between the Debye-Hückel potential obtained via a numerical approach (NA) and the analytical Debye-Hückel potential obtained under the Yukawa approximation (YA), as labelled. Bottom: comparison between the Yukawa-like potential (YA) and the coarse-grained potential (CG), as labelled

$Z_{p1} \equiv Z_{p2} \equiv Z_p = 90$; lengths are given in units of the particle diameter (i.e., $2\sigma_c$), while charges are given in units of the elementary charge. The considered IPCs are neutral (i.e., $Z_{tot} = Z_c + 2Z_p = 0$). We note that the relative value between the charges affects the symmetries of the effective potential, while the effect of their absolute value is ruled out by the normalization: the pair potential is indeed normalized by the value of the minimum energy, i.e., by the contact energy between two IPCs in the EP configuration. Having set these parameters, we can derive the effective potential between two IPCs within the Debye-Hückel framework, as described in Sect. 2.2. The top panels of Fig. 2.5 report the comparison between the numerical approach and the Yukawa-like approximation: the agreement between the two potentials is rather good both for the radial (left panel) and angular (right panel) behavior.

With the Debye-Hückel potential at hand, we can derive the coarse-grained parameters for our patchy model. We first fix the interaction range to $\delta = 0.2$ (corresponding to $\kappa\delta = 2$) and the patch size to $\gamma_1 \equiv \gamma_2 \equiv \gamma \approx 45°$ (corresponding to $a = 0.22$, with $\sigma_{p1} \equiv \sigma_{p2} \equiv \sigma_p = 0.38$). We then perform the "max" mapping with respect to the Yukawa-like pair potential to derive the energy strengths of the characteristic reference configurations (EE, EP, and PP). We obtain $u_{EE} = 0.2457$, $u_{EP} = -3.1169$, and $u_{PP} = 21.2298$, giving rise to the following contact energies $e_{EP} = -1$ (by construction), $e_{EE} = 0.1173$, and $e_{PP} = 4.0806$. The bottom panels of Fig. 2.5 report the comparison between the Yukawa-like and the coarse-grained potential: the agreement between the two is rather good both for the radial (left panel) and angular (right panel) behavior, and fully justifies the use of the simpler coarse-grained potential (vs the Debye-Hückel one) in many body simulations.

2.5   Conclusions

Colloids with patterned surfaces, commonly referred to as patchy particles, are considered as very versatile building entities whose shape and surface decoration can be designed such that they support the self-assembly of target structures with desired properties. The self-organizing behavior of these units is based on their ability to form directional and highly selective bonds via the specific interaction features of the different surface areas. The versatility of this class of particles can be even enhanced if their surface decoration is characterized by charged regions. Charged spherical colloids decorated by oppositely charged regions can be referred to as inverse patchy colloids (IPCs).

In this chapter, we have described in detail a coarse-graining path to obtain the effective interaction potential between IPCs with rich surface patterns. The coarse-grained description of the pair potential between IPCs is derived by combining a mean field theoretical approach with a simple patchy model.

The theoretical description treats the electrolytic solution as a suspension of impenetrable, charged colloidal particles in a liquid dielectric solvent containing co- and counter-ions. For a linear, homogeneous, and isotropic dielectric medium, the Maxwell equations of electrostatics lead (in the absence of external fields) to a differential equation of Poisson type, which relates the electrostatic potential of the system with the ionic charge density [27]. In a mean field approach, the equilibrium charge density obeys Boltzmann statistics; thus, the resulting equation that describes the electrostatic potential, known in literature as the Poisson-Boltzmann equation, is a non-linear differential equation. The linearized Poisson-Boltzmann approach is referred to as the Debye-Hückel description [28]. It is worth noting that, although strictly suitable only for the dilute case, the linear approximation provides surprisingly good results also for denser systems [29]. The extensions to concentrated solutions often require the replacement of the bare colloidal charge with a so-called effective charge, which takes into account the strong condensation of counterions on the Stern-layer around the colloids [30]. At any rate, the functional form of the Debye-Hückel potential can be preserved and it provides for a realistic description of experimental data for a vast variety of physical situations. It should also be noted that deviations from the Debye-Hückel theory occur not only at high ion concentrations and but also for multivalent electrolytes: the linear approximation is valid only for symmetric electrolytes [31]; when multivalent—especially asymmetric—electrolytes are involved, the Debye-Hückel theory is not able to take into account the ion–ion correlations and thus different approaches are needed [32]. For homogeneously charged spherical colloids, the Debye-Hückel approach leads to the traditional Derjaguin-Landau-Verwey-Overbeek pair interaction [22]. The same approach can be used for heterogeneously charged colloids [1, 21]: first an expression of the screened electrostatic potential generated by a single IPC is obtained within the Debye-Hückel approach and then, based on this information, the effective pair potential between a pair of IPCs is built. We note here that, in our derivation of the effective pair interaction between IPCs, we have neglected—for the sake of simplicity—that colloids typically have a dielectric constant different from the one of the surrounding solvent: this simplification can of course be removed [25]. When such a dielectric discontinuity is taken into account, one should also take care of possible effective interactions arising from inhomogeneous polarization charge distributions [33].

The simple patchy model is designed to reproduce the symmetries of the theoretical description. It features a hard spherical particle carrying a fixed number of interaction sites placed in a well-defined geometry inside the colloid. As a consequence of the site positions, their corresponding interaction spheres extend partially outside the hard core particle, defining in this way the patches. The model is characterized by three sets of independent parameters: the interaction ranges of the different surface regions, their surface extents, and their interaction strengths. Since the characteristic interaction distances are determined by the electrostatic screening of the surrounding solvent, all entities of the colloid are assumed to have the same interaction range, irrespective of the surface regions involved in the interaction. The patch sizes are ideally determined by the corresponding feature of experimentally synthesized particles. Finally, the energy parameters of the model are related to the charges involved in the interactions. These charges are responsible for the ratio between the attractive and repulsive contributions to the pair energy associated to the different (patch/patch, patch/non-patch, and non-patch/non-patch) interactions.

The specific form of the pair potential is based on the postulate that each of these contributions can be factorized into an energy strength and a geometrical weight factor, the latter one being given by the distance dependent overlap volume of the involved interaction spheres.

In order to provide the simple patchy model with parameters that are directly related to the physical quantities of the underlying microscopic system, the available Debye-Hückel description can be taken advantage of. The procedure to determine the interaction parameters of the model is mainly based on a mapping from the Debye-Hückel to the patchy potential: such a mapping relies on the identification of the most representative particle–particle configurations. The selection of these reference configurations strongly depends on the details of the particle design and must be carefully reasoned and tested. The resulting coarse-grained description is advantageous in many body simulations because the model pair energy is the sum over simple products of geometric (orientation- and distance-dependent) factors and energy factors, associated to the different ways of interaction that characterize the IPC. The proposed method allows to study the self-assembly of heterogeneously charged colloidal systems that are observed in experiments [2].

It is worth noting that, besides the possibility to experimentally synthesize IPC model units, many colloidal systems are composed of particles with a heterogeneous surface charge. Indeed, when dispersed in a microscopic medium, colloids can acquire a possibly inhomogeneous surface charge due to dissociation of surface groups and/or preferential adsorption of charged species [23]. Moreover, also naturally occurring systems such as proteins and virus capsids are known to have heterogeneously charged surfaces which give rise to specific collective behaviors [34–39]. Thus, appropriately developed IPC models could also be used to gain a better insight into the assembly of these natural systems.

Acknowledgements I am indebted to Christos N. Likos and Gerhard Kahl for their valuable contributions to the modeling of heterogeneously charged colloids. I gratefully acknowledge the Alexander von Humboldt Foundation for financial support through a Research Fellowship, and the Austrian Science Fund (FWF) for financial support under Proj. Nos. M1170-N16 (Lise Meitner Fellowship) and V249-N27 (Elise Richter Fellowship).

References

1. Bianchi E, Kahl G, Likos CN. Inverse patchy colloids: from microscopic description to mesoscopic coarse-graining. Soft Matter. 2011;7:8313.
2. van Oostrum PDJ, Hejazifar M, Niedermayer C, Reimhult E. Simple method for the synthesis of inverse patchy colloids. J Phys Condens Matter. 2015;27:234105.
3. Kalyuzhnyi YV, Vasilyev OA, Cummings PT. Inverse patchy colloids with two and three patches. Analytical and numerical study. J Chem Phys. 2015;143:044904.
4. Yigit C, Heyda J, Dzubiella J. Charged patchy particle models in explicit salt: ion distributions, electrostatic potentials, and effective interactions. J Chem Phys. 2015;143:064904.
5. Blanco MA, Shen VK. Effect of the surface charge distribution on the fluid phase behavior of charged colloids and proteins. J Chem Phys. 2016;145:155102.

6. Hieronimus R, Raschke S, Heuer A. How to model the interaction of charged Janus particles. J Chem Phys. 2016;145:064303.
7. Bharti B, Kogler F, Hall CK, Klapp SHL, Velev OD. Multidirectional colloidal assembly in concurrent electric and magnetic fields. Soft Matter. 2016;12:7747.
8. Dempster JM, de la Cruz MO. Aggregation of heterogeneously charged colloids. ACS Nano. 2016;10:5909.
9. Pawar A, Kretzschmar I. Fabrication, assembly, and application of patchy particles. Macromol Rapid Commun. 2010;31:150.
10. Bianchi E, Blaak R, Likos CN. Patchy colloids: state of the art and perspectives. Phys Chem Chem Phys. 2011;13:6397.
11. Wang Y, Wang Y, Breed DR, Manoharan VN, Feng L, Hollingsworth AD, Weck M, Pine DJ. Colloids with valence and specific directional bonding. Nature 2012;491:51.
12. Chen Q, Bae SC, Granick S. Directed self-assembly of a colloidal kagome lattice. Nature 2011;469:381.
13. Likos CN, Blaak R, Wynveen A. Computer simulations of polyelectrolyte stars and brushes. J Phys Condens Matter. 2008;20:494221.
14. Božič AL, Podgornik R. Symmetry effects in electrostatic interactions between two arbitrarily charged spherical shells in the Debye-Hückel approximation. J Chem Phys. 2013;138:074902.
15. Bianchi E, Likos CN, Kahl G. Self-assembly of heterogeneously charged particles under confinement. ACS Nano. 2013;7:4657.
16. Bianchi E, Likos CN, Kahl G. Tunable assembly of heterogeneously charged colloids. Nano Lett. 2014;14:3412.
17. Noya EG, Kolovos I, Doppelbauer G, Kahl G, Bianchi E. Phase diagram of inverse patchy colloids assembling into an equilibrium laminar phase. Soft Matter 2014;10:8464.
18. Noya EG, Bianchi E. Phase behaviour of inverse patchy colloids: effect of the model parameters. J Phys Condens Matter. 2015;27:234103.
19. Ferrari S, Bianchi E, Kahl G. Spontaneous assembly of a hybrid crystal-liquid phase in inverse patchy colloid systems. Nanoscale 2017;9:1956.
20. Bianchi E, van Oostrum PDJ, Likos CN, Kahl G. Inverse patchy colloids: synthesis, modeling and self-organization. Curr Opin Colloid Interface Sci. 2017;30:8.
21. Stipsitz M, Bianchi E, Kahl G. Generalized inverse patchy colloid model. J Chem Phys. 2015;142:114905.
22. Verwey EJW, Overbeek JThG. Theory of the stability of lyophobic colloids. Amsterdam: Elsevier; 1948.
23. Russel WB, Saville DA, Schowalter WR. Colloidal dispersions. Cambridge: Cambridge University Press; 1989.
24. Abramowitz M, Stegun IA. Handbook of mathematical functions: with formulas, graphs, and mathematical tables. New York: Dover; 1965.
25. Hoffmann N, Likos CN, Hansen J-P. Linear screening of the electrostatic potential around spherical particles with non-spherical charge patterns. Mol Phys. 2004;102:857.
26. Frenkel D, Smit B. Understanding molecular simulations. San Diego: Academic; 2002.
27. Jackson JD. Classical electrodynamics. 3rd ed. New York: Wiley; 1999.
28. Debye P, Hückel E. The theory of electrolytes. I. Lowering of freezing point and related phenomena. Phys Z. 1923;24:185.
29. El Masri D, van Oostrum PDJ, Smallenburg F, Vissers T, Imhof A, Dijkstra M, van Blaaderen A. Measuring colloidal forces from particle position deviations inside an optical trap. Soft Matter. 2011;7:3462.
30. Trizac E, Bocquet L, Aubouy M, von Grünberg HH. Alexander's prescription for colloidal charge renormalization. Langmuir 2003;19:4027.
31. Guldbrand L, Jönsson B, Wennerström H, Linse P. Electrical double layer forces. A Monte Carlo study. J Phys Condens Matter. 1984;80:2221.
32. dos Santos AP, Diehl A, Levin Y. Colloidal charge renormalization in suspensions containing multivalent electrolyte. J Phys Condens Matter. 2000;132:104105.

33. Barros K, Luijten E. Dielectric effects in the self-assembly of binary colloidal aggregates. Phys Rev Lett. 2014;113:017801.
34. Daniel MC, Tsvetkova IB, Quinkert ZT, Murali A, De M, Rotello VM, Kao CC, Dragnea B. Role of surface charge density in nanoparticle-templated assembly of bromovirus protein cages. ACS Nano. 2010;4:3853.
35. Gögelein C, Nägele G, Tuinier R, Gibaud T, Stradner A, Schurtenberger P. A simple patchy colloid model for the phase behavior of lysozyme dispersions. J Chem Phys. 2008;129:085102.
36. Božič AL, Šiber A, Podgornik R. How simple can a model of an empty viral capsid be? Charge distributions in viral capsids. J Biol Phys. 2012;38:657.
37. Kurut A, Persson BA, Åkesson T, Forsman J, Lund M. Anisotropic interactions in protein mixtures: Self assembly and phase behavior in aqueous solution. J Phys Chem Lett. 2012;3:731.
38. Roosen-Runge F, Zhang F, Schreiber F, Roth R. Ion-activated attractive patches as a mechanism for controlled protein interactions. Sci Rep. 2014;4:7016.
39. Li W, Persson BA, Morin M, Behrens MA, Lund M, Oskolkova MZ. Charge-induced patchy attractions between proteins. J Phys Chem B. 2015;119:503.

A Nucleotide-Level Computational Approach to DNA-Based Materials

3

Flavio Romano and Lorenzo Rovigatti

3.1 Introduction

DNA has been dubbed "the molecule of life" for its pivotal role in biology. It encodes the genetic information of all living beings in the linear arrangement of four different "letters" along a polymeric backbone. These letters, the four nucleobases adenine (A), thymine (T), cytosine (C) and guanine (G), bind selectively with each other, A with T and C with G [1]. This pairing mechanism, named after Watson and Crick, together with the chirality of the DNA molecule which encodes a direction along the polymer, provides a way of storing information in a unidimensional structure [2]. Other interactions, such as non-Watson–Crick pairing, are also present and can stabilise exotic structures such as G-quadruplexes [3, 4], but the Watson–Crick pairing is responsible for the vast majority of the secondary structure of DNA and thus for its typical behaviour.

The high specificity of the Watson–Crick pairing is essential in vivo to accurately replicate genetic code. However, it also makes nucleic acids amenable to be artificially exploited as building blocks for nano- and mesoscopic applications. The notion of using DNA for technological uses dates back to the 1980s, when Seeman laid the foundations of a new field, DNA nanotechnology [5, 6]. The outstanding development of a theoretical understanding of DNA self-assembly and of DNA synthesis, manipulation and characterisation has made it possible to manufacture all-DNA molecular motors [7], logic gates [8] and finite-sized objects with pre-designed shapes such as polyhedra [9, 10], tubes [11, 12] or even more complicated

F. Romano

Dipartimento di Scienze Molecolari e Nanosistemi, Università Ca' Foscari, Via Torino 155, Venezia, Italy

L. Rovigatti (✉)
CNR-ISC, Uos Sapienza, Piazzale A. Moro 2, 00185 Roma, Italy
e-mail: lorenzo.rovigatti@gmail.com

© Springer International Publishing AG, part of Springer Nature 2017
I. Coluzza (ed.), *Design of Self-Assembling Materials*,
https://doi.org/10.1007/978-3-319-71578-0_3

structures such as DNA origami [13]. These structures, which reliably assemble in the designed shape, are commonly used as nano-scaffolds for high precision experiments [14, 15] or even as drug delivery vectors [16]. On the materials science side, early results obtained by Seeman and others demonstrated that DNA can also be employed as a building block for the generation of soft-matter, ordered and disordered bulk phases [17–20]. There are two general ways of using DNA in this context. Firstly, DNA can be used as a link to bridge nano- or micrometer sized particles. This idea, which revolves around the concept of grafting DNA strands onto the surface of colloidal particles, has been exploited to generate a wide range of crystalline and non-crystalline phases [17,21–23]. In these systems the selectivity of DNA is used to tune the effective interaction between the main components of the material, e.g. colloidal particles. By contrast, DNA can also be used on its own to generate so-called all-DNA materials. These materials are mainly (if not entirely) made of DNA. Both disordered and ordered materials have been synthesised this way [19,24].

In the last few years the significant progress achieved by the fast-moving field of DNA nanotechnology has spurred the development of new theoretical and computational tools to complement experimental work. However, the time, ranging from nanoseconds to hours, and length scales, ranging from nanometers to almost millimiters, involved in these processes are very diverse and thus require different approaches. For example, in numerical simulations the characteristic length scale of the chosen model puts a constraint, due to the limited amount of available computer power, on the maximum length and time scales that can be investigated. Therefore, it is important to choose the right model in relation to the system to be studied: the larger the characteristic length, the more coarse-grained the model needs to be. In this chapter we will focus on soft-matter systems composed by particles whose sizes range from tens of nanometers to micrometers. If compared to the size of a nucleotide, these length scales are orders of magnitude larger, suggesting to employ models that describe DNA on a rather coarse-grained level. In fact, a substantial effort has been spent in the development of highly coarse-grained models that take into account DNA–DNA interactions in colloidal systems in an *effective* way [25–28]. However, the heterogeneous nature of DNA is such that the global properties of a whole system may depend on the identity or conformation of one or a few nucleotides, thereby making it mandatory to use a highly detailed model, regardless of the other characteristic length scales of the system at hand. This class of systems, which encompasses the majority of all-DNA materials, requires a *multiscale* approach.

This chapter is organised as follows. In Sect. 3.2 we will discuss the most important properties of DNA in relation to its self-assembly ability and recent work on DNA in materials science. Section 3.3 will give an overview of the computational models used to simulate DNA systems. In Sect. 3.4 we will focus on the implications of using a coarse-grained nucleotide-level DNA model, oxDNA, for the investigation of bulk all-DNA materials. Section 3.5 will provide some perspectives for future work in the field.

3.2   Modelling DNA

3.2.1   The DNA Molecule

DNA (deoxyribonucleic acid) is a biological macromolecule composed of units called nucleotides. Each nucleotide is made of a phosphate, a sugar and a nucleobase, either A, C, G or T. Nucleotides are linked together by covalent bonds which join the sugar and the phosphate of consecutive nucleotides. The most common form of double-stranded DNA (dsDNA) is B-DNA, which is a right-handed double helix of diameter ≈ 2.0 nm and pitch ≈ 3.3 nm (roughly 10 base pairs). The helix is stabilised not only by hydrogen bonds between bases, but also by stacking interactions among the aromatic rings of consecutive nucleotides. These stacking interactions are largely responsible for the high stiffness of the double helix, which has a persistence length of ≈ 150 base pairs [29]. By contrast, single-stranded DNA (ssDNA) has a persistence length of a few bases. This large difference in stiffness between the two conformations is extremely important in applications, since it allows to produce hierarchical structures with tunable flexibility by employing rigid parts made of dsDNA joined together by flexible ssDNA linkers.

The structural and mechanical properties of DNA are highly dependent on external conditions. At high enough temperature T DNA is always in a single-stranded conformation. As T is lowered, base pairing becomes important and strands begin to hybridise, leading to the formation of secondary structure. The single- to double-stranded transition is identified by a melting temperature, defined as the temperature T_m at which a system is half of the time in a conformation and half of the time in the other one, and by a transition width, which is a measure of the temperature range over which the transition occurs. Hybridisation of DNA duplexes is a cooperative transition, and as such its associated T_m grows as the length of the strands increases. Temperature also affects, although to a lesser extent, the behaviour of single-stranded DNA having sequences that minimise secondary structure, such as poly(dA), which, at low T, exhibits stacked regions with mechanical properties intermediate between those of dsDNA and unstacked ssDNA.

Since DNA is a negatively charged macromolecule, its thermodynamic and dynamic behaviours, and to a lesser extent its structure, are affected by the ionic strength of the solution, usually measured in terms of the salt concentration c. Indeed, at high salt concentration ($c \geq 0.5$ M [Na^+]), at which most DNA nanotechnology experiments are carried out, electrostatic interactions are mostly screened out and the extent of the effective repulsion between DNA strands due to charges is greatly reduced with respect to physiological conditions ($c \approx 0.15$ M [Na^+]). Consequently, melting temperatures and association rates increase. In addition, the decrease of the DNA–DNA repulsion affects the conformation of junctions and of other motifs in regions of high local concentration. As a last point, we note that DNA conformation and mechanical properties, such as persistence length, are also sensitive to the valency of the salt in solution. For example, it has been shown that multivalent counterions can condensate dsDNA [30].

Different conditions of pH also affect DNA. In particular, it is known that the strength of the inter-base hydrogen bonds decreases as the pH value departs from physiologic conditions (pH $\simeq$ 7.4) [31]. As a result, duplex melting temperatures drop and the DNA mechanical properties change for pH values far from 7.4 [32–34].

3.2.2 Exploiting DNA in Artificial Applications

The Watson–Crick pairing mechanism makes DNA highly selective: it is possible to design DNA sequences that, under the right external conditions, bind to other molecules or to specific parts of other molecules with high precision. This selectivity is not limited to DNA–DNA interactions, but extends to DNA–RNA and DNA–protein systems as well. In a biological environment, this molecular recognition mechanism is required by the cell nanomachinery that takes care of gene expression, damage repairing and genome replication. However, the specificity of DNA interactions can also be exploited in artificial applications. Indeed, strands with carefully designed sequences can self-assemble in a reliable fashion to generate predetermined complexes that interact in a controlled way, both among themselves and with other objects that may be present in solution. The interaction between double-stranded parts of the structure can be enforced through the use of *sticky ends*, i.e. short single-stranded overhangs that are designed to specifically bind with each other. The size of all-DNA objects generated in this way can exceed one hundred nanometers, while their internal structure can be finely tuned at the level of single nucleotides, i.e. nanometers. Thanks to the high persistence length of dsDNA (which can be further increased by using more complicated motifs [35]), these nanostructures retain an average shape that can be exploited in applications [16, 36], even though some flexibility is unavoidable [29].

Another advantage of DNA lies in its availability. The price of viral and synthetic DNA dropped down considerably in the last decade, and it is now possible to obtain large quantities of strands of synthetic DNA with the desired sequences. Many enzymes, developed by nature to work with DNA in the cell, are commercially available and can be used in experiments to manipulate DNA, e.g. to cut, join or repair DNA strands. DNA and enzymes can also be exploited to selectively replicate in vitro some of the processes happening in vivo. For example, it is possible to design strands that self-assemble into an all-DNA hydrogel which, by employing part of the cell machinery can be used to express large amounts of proteins [37]. The possibility of using synthetic DNA, as well as RNA, in biological contexts is an important feature of nanotechnology based on nucleic acids. In addition to the aforementioned cell-free synthesis of proteins, DNA has been used to design and produce nanostructures that actively react in the presence of specific biomolecules or in particular environments. For example, it is already possible to fabricate nanoscale robots that can store and selectively release a molecular payload in vivo [16, 38].

The specificity of DNA can also be exploited in materials science to introduce a tunable binding mechanism between the basic constituents. DNA-based soft-matter building blocks have been historically divided into two distinct categories: DNA-coated colloids DNA-CCs and all-DNA supramolecular constructs (all-DNA). Both strategies exploit the selective binding provided by the Watson–Crick mechanism to introduce an effective, temperature-dependent interaction between the basic constituents of the system, although the two schemes are essentially different and have different advantages and disadvantages. We note that small nanoparticles grafted with very sparse ultrashort strands, such as the system investigated in [39], can be seen as intermediate between DNA-CCs and all-DNA systems.

3.2.2.1 DNA-Coated Colloids

The possibility of tuning the mutual interactions between nano- and microsized particles by functionalising their surface makes them suitable for many technological and medical applications [23, 40]. For example, polymers have been used for decades to sterically stabilise colloidal solutions [41]. In addition to providing a tunable repulsion, grafted DNA can also be used to add a controllable *attraction* between colloids. By carefully choosing the strand sequences and grafting density, DNA-CCs have been used to create non-compact crystals [17, 42], crystals with tunable lattice parameters [21], gels [22] and more [23, 40]. Most of these results have been obtained by using a uniform DNA-grafting density. There are a few notable exceptions. Feng et al., for example, have developed a simple method to make micron-sized DNA-CCs with a single patch and a very high yield [43]. By contrast, Wang et al. use a much more versatile, albeit laborious, technique to create "colloidal analogues of atoms with valence" by selectively grafting DNA-strands on the protrusions of small clusters of amidinated polysterene microspheres with well-defined symmetries [44]. A similar strategy, which employs DNA-grafted polyhedra blocks to add directional binding, has been recently proposed [45]. It is also possible to place single DNA strands with specific arrangements on the surface of colloids [46, 47]. Halverson and Tkachenko have numerically shown that these directionally functionalised DNA-CCs could provide a route to generate error-free, mesoscopic structures with high yield [48].

In general, all the aforementioned methods can be used to synthesise colloids that can, and in some cases have been used to, self-assemble into crystalline phases. However, the steepness of the DNA melting curves makes it very difficult to generate the defect-free, large crystals that are required for applications. Indeed, the experimental conditions have to be chosen and controlled very carefully so as to avoid falling off of equilibrium, which would otherwise lead to the formation of either disordered aggregates or highly defective crystals. An effort to overcome such limitation led to the development of a method to synthesise DNA-CCs with mobile linkers [49]. This technique makes it possible for DNA strands to diffuse over the colloid surface, allowing particles to anneal to the equilibrium state much faster than regular DNA-CCs. Angioletti-Uberti et al. have built on this idea and suggested to exploit the many-body nature of the interaction between DNA-CCs with mobile linkers by grafting additional non-binding DNA strands. These can thus exert a non-specific repulsion that, depending on external parameters such as temperature or salt concentration, sets the maximum number of bonds that each particle can establish.

Simulation results confirm that, in contrast with DNA-CCs with immobile DNA linkers, these particles can self-assemble into open structures [50].

More recently, the issue of annealing DNA systems has seen the breakthrough development by Rogers et al., who devised a strategy to decouple the final self-assembly product from its pathway [51]. Their idea, which revolves around concepts borrowed from the DNA nanotechnology field, can be applied to any DNA-based system to synthesise systems undergoing robust crystallisation, solid–solid transitions, reentrant melting, reentrant gelation and self-healing [51–53].

3.2.2.2 All-DNA Constructs

DNA constructs are fixed-sized objects composed by a certain number of single strands that are designed to self-assemble into specific structures. Bonding between different constructs can be provided by blunt dsDNA-dsDNA stacking or by short unpaired overhangs, called *sticky ends*, placed at the end of some (or all) of the strands. The number of these sticky ends is fixed by design and controls the maximum number of possible bonds that each construct can establish. This built-in *valence* makes DNA constructs ideal candidates to synthesise materials that have been observed in toy models of limited-valence colloidal particles, such as equilibrium gels and open crystals [54–56]. However, these constructs require an accurate synthesis and long purification and annealing protocols [19]. Generally, all-DNA materials require a multi-step (hierarchical) self-assembly process: first, single strands assemble into DNA-constructs or tiles. These constructs, in turn, self-assemble into larger objects or structures upon further lowering the temperature [5]. This strategy has been employed in the past to generate crystalline structures of different types [5, 24]. More recently, DNA has been incorporated as a tool to investigate and generate soft-matter disordered or partially ordered materials. For example, Nakata and co-workers have shown that ultrashort DNA strands can stack onto each other and self-assemble into long chains that, at high concentrations, form liquid crystalline states [57]. The dependence on the sequence, type of nucleic acid (DNA or RNA) and the effect of the presence of dangling ends have all been investigated experimentally [57–59], showing that the final state is deeply affected by even small changes in the building blocks. A fundamental understanding of the self-assembly of these systems has been provided by the theory developed by De Michele and co-workers. This theory, together with accompanying numerical simulations, has shown that the formation of these all-DNA liquid crystals can be understood by modelling double strands as cylinders with two attractive patches, resulting in semi-quantitative agreement with experiments [60–62].

The connection between simple toy models of anisotropic particles and DNA-constructs has been further strengthened by recent experiments on trivalent and tetravalent DNA nanostars, i.e. DNA constructs with three and four arms, respectively. Biffi and co-workers have carried out measurements of the low-density phase diagram and dynamics of these nanostars, providing the first experimental confirmation of the dependence of the size of the gas-liquid instability region on the valence [19], which was theorised for the first time a decade ago [54]. The dynamics in the equilibrium gel of DNA nanostars has also been characterised [63,64]. On the numerical side, simulations of a coarse-grained model have been shown to match quantitatively experimental results [65], supporting the experimentally observed thermodynamic stability of the disordered gel with respect to crystallisation [66], in line with recent results on toy models of patchy particles [67].

3.3 Coarse-Graining DNA

Investigating DNA by means of theory or simulation requires a careful choice of the level of description, which depends on the physics relevant for the question to be addressed. This is not straightforward in our case, since DNA is involved in processes that span length scales that go from nanometers (single bases) to micrometers (DNA-coated colloids) or even millimiters (chromosomes). Figure 3.1 shows some representative systems. For this reason, many approaches have been developed to describe DNA at the different levels of detail, and possibly the greatest challenge for the future of the field is to find a satisfactory way to develop reliable multi-scale approaches.

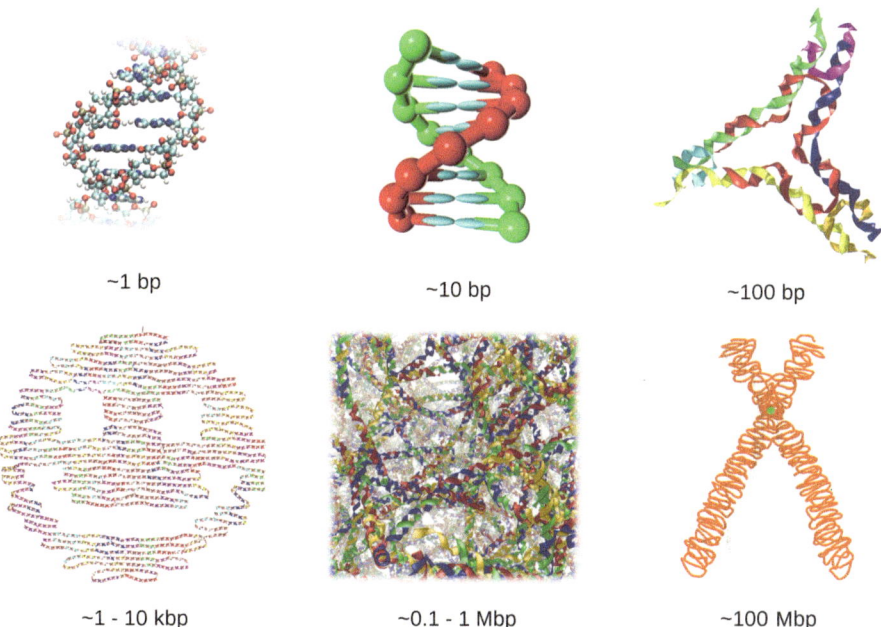

~1 bp   ~10 bp   ~100 bp

~1 - 10 kbp   ~0.1 - 1 Mbp   ~100 Mbp

Fig. 3.1 The characteristic length scales of DNA-based systems, and hence the minimal resolution required to describe them, can vary considerably. Here we show sketches of a few selected systems and indicate their typical size, measured in bp (number of base pairs). Top row, from left to right: the atomic structure of a few base pairs, a short duplex and a DNA tile, which can hierarchically self-assemble into higher-order objects [10]. Bottom row, from left to right: a DNA origami [13], an all-DNA gel [65] and a chromosome (adapted from a picture by Manske)

Long (>100 nm) length scales require a very coarse-grained description, and in this context (duplex) DNA is usually modelled as a worm-like polymer. This level of description, which does not take into account single nucleotides, is useful to investigate the effects of the elastic properties (e.g. the twisting and bending moduli) of long strands of DNA, often in a biological environment [68, 69]. However, it is not apt to describe smaller DNA fragments or processes such as hybridisation or bubble formation. Conversely, extremely detailed methods, which make use of quantum chemistry techniques, can be used to assess the interactions between nearest-neighbour nucleotides in vacuum [70, 71]. Larger systems can be investigated by means of all-atom simulations, in which DNA and solvent atoms are modelled as point-like particles interacting through effective potentials. Thanks to the staggering increase of computer power, it is possible to employ this level of detail to simulate large structures, even though the time scales accessible do not exceed the microsecond, thereby making the simulation of rare events a challenge [72, 73]. In addition, the parameters of these models are fitted to specific conformations, usually the DNA duplex, and it is currently not known whether using them in different contexts, such as duplex formation or for single strands, yields realistic results.

In between the all-atom and polymer descriptions are models that describe DNA at the nucleotide level. The simplest coarse-grained models of this kind are statistical [74, 75]. They provide an estimate of the free energy gained by forming base pairs as a function of strand sequence and solution conditions, such as the salt concentration. Therefore, these models can be used to predict the thermodynamics, i.e. stability and melting curves, of any secondary structure. However, this level of description completely neglects the three-dimensional structure, kinetics and mechanics of DNA. Moreover, it can describe only the simplest non-trivial secondary structures such as single-stranded loops and the most simple pseudoknots.

These issues can be overcome, at least nominally, by employing computational models that represent nucleotides as a collection of interacting sites. The larger-than-ever availability of experimental data to fit has triggered the development of many coarse-grained models that describe DNA at this level of detail, in which most of the degrees of freedom of the atoms composing DNA are integrated out. Depending on the adopted coarse-graining procedure, the resulting model can be classified into one of two broad "families". In *bottom-up* approaches the parameters of the model are fitted to results (usually structural observables such as local correlations) obtained from simulations of finer-grained models, such as all-atom force fields. The bottom-up coarse-grained models tend to perform very well when used to study the structural properties of dsDNA, reflecting the fact that the all-atom models have been fitted to reproduce experimental structures [76–79]. However, it is not currently known how well the behaviour of single strands, as well as melting and hybridisation, is described by atomistic simulations and hence, consequently, by bottom-up models.

A complementary approach is exemplified by the so-called top-down family of models, which are parametrised after global properties of DNA, such as duplex melting temperatures or persistence length, and feature force-fields with heuristic interactions. These models tend to yield a worse description of the local structure of DNA but can be used with more confidence to investigate larger-scale phenomena such as hybridisation, hairpin formation, stacking transition, mechanical folding, strand displacement and more [80–84]. At this level of detail the solvent is often, but not always, treated *implicitly*, i.e. its effects are incorporated in the effective interactions between the sites [85, 86]. Although the two approaches have different strengths and weaknesses, a point that is often overlooked is that in both bottom-up and top-down approaches representability problems are inevitable [87] since the number of degrees of freedom is vastly reduced with respect to that of the real system.

3.4   Insight on DNA-Based Materials from Nucleotide-Level Simulations

The coarse-grained models employed for the investigation of soft-matter systems often have characteristic lengths comparable with the largest particle sizes, and the degrees of freedom of the smaller components, from solvent molecules to co-solutes, are integrated out and accounted for with effective interactions. This coarse graining can be performed in different ways, depending on the quality of the agreement one wants to achieve. For example, the behaviour of DNA-CCs can be qualitatively captured by simple models such as the square-well model [22], but a more quantitative description requires a careful consideration of the role of the DNA linkers. If the DNA length is smaller but comparable with the colloid diameter one can model the DNA strands as Gaussian polymeric chains and implement ad-hoc Monte Carlo moves that take into account hybridisation [88]. The DNA-mediated interaction between colloids can also be parametrised by considering the free-energy difference due to the hybridisation of DNA strands grafted onto two planar surfaces and using the Derjaguin approximation [28]. In all these coarse-grained approaches the hybridisation free energy is usually evaluated by using a statistical model such as SantaLucia's nearest-neighbour model [74]. These approaches neglect the helical structure and the majority of the internal degrees of freedom and of DNA, as they usually do not play a significant role in the processes happening at the colloid length scale. However, when the basic constituents of the system are mainly (or exclusively) composed of DNA, as it is the case for all-DNA materials [61,65,89], or when the strands undergo processes more complicated than simple duplex hybridisation, such as hairpin formation or toe-hold mediated strand displacements [21,89], it is useful to use models that describe DNA at a finer level of detail. In what follows we will demonstrate that a top-down model describing DNA at the nucleotide level can be used to both design and gather insight on the self-assembly of DNA-based materials. In particular, we will focus on DNA liquid crystals and DNA gels, which have been investigated with oxDNA, a coarse-grained DNA model that has been designed for simulating DNA in the context of DNA nanotechnology [84].

3.4.1 The OxDNA Model

OxDNA represents DNA as a string of nucleotides, where each nucleotide (sugar, phosphate and base group) is a rigid body with interaction sites for backbone, stacking and hydrogen-bonding interactions. The potential energy of the system is

$$V_0 = \sum_{\langle ij \rangle} \left(V_{\text{b.b.}} + V_{\text{stack}} + V'_{\text{exc}} \right) + \sum_{i,j \notin \langle ij \rangle} \left(V_{\text{HB}} + V_{\text{cr.st.}} + V_{\text{exc}} + V_{\text{cx.st.}} \right), \qquad (3.1)$$

where the first sum is taken over all nucleotides that are nearest neighbours on the same strand and the second sum comprises all remaining pairs. The explicit forms of the interactions between nucleotides can be found in [90] and [91]. The hydrogen bonding (V_{HB}), cross stacking ($V_{\text{cr.st.}}$), coaxial stacking ($V_{\text{cx.st.}}$) and stacking interactions (V_{stack}) explicitly depend on the relative orientations of the nucleotides as well as on the distance between interaction sites. The backbone potential $V_{\text{b.b.}}$ is an isotropic finitely extensible spring that imposes a finite maximum and minimum distance between neighbours, mimicking the covalent bonds along the strand. The coaxial stacking term is designed to capture stacking interactions between non-neighbouring bases, usually on different strands. All interaction sites also have isotropic excluded volume interactions V_{exc} or V'_{exc}. The values of the parameters entering into the functional forms of the interaction potentials have been chosen so as to yield the right thermodynamics (as given by the nearest-neighbour model of SantaLucia [74]), as well as the structure and mechanics of both ssDNA and dsDNA. We stress that the difference in the mechanical properties of the two conformations, which oxDNA is designed to exhibit, is of utmost importance for nanotechnology applications.

In the first implementation of oxDNA there were no sequence-dependent parameters (apart from the built-in selectivity of the Watson–Crick base pairing), no minor and major grooving and the model was parametrised at a fixed salt concentration (0.5 M [Na^+]). OxDNA has since been extended to provide sequence-dependence for the V_{stack} and V_{HB} terms [92], a more detailed helical structure and salt dependence [85]. An RNA model based on oxDNA has also been developed [93].

OxDNA has been designed to investigate processes happening at the nanoscale. It has been used to understand basic DNA properties, such as hybridisation [94], hairpin formation [95], plectoneme formation [96], as well as processes pertinent to DNA nanotechnology, such as strand displacement [97] and the assembly of a small DNA origami [98].

In what follows we will show that it is possible to employ such a detailed model to quantitatively investigate the phase behaviour of all-DNA materials. This can be done either by performing simulations of bulk systems or by computing quantities that can be plugged into more coarse-grained numerical or theoretical approaches.

3.4.2 Direct Simulations

DNA-based building blocks are complex, heterogeneous objects. Their mutual interactions are often non-trivial, and multi-body effects are common [99]. Therefore, understanding the effect of the shape and properties of a DNA construct on the behaviour of the resulting bulk material is not a simple task. Although algorithm and hardware advances have made it technically possible to simulate systems made of tens of thousands bases with nucleotide-level models such as oxDNA, the direct simulation of an all-DNA material remains a cumbersome task and requires a substantial numerical effort. The reason is twofold. First of all, and this applies to the majority of DNA models, oxDNA features highly directional interactions which are numerically demanding and affect performance. As a result, it takes much more computer power to simulate N nucleotides than N particles interacting through a simple potential such as the Lennard–Jones interaction [100]. Moreover, hybridisation is a slow process, and simulating the breaking and formation of dsDNA in unconstrained simulations takes a long time. Secondly, since constructs are composed of many nucleotides, the total number of constructs that can be simulated is usually much lower than the number of particles in coarse-grained systems. Unpleasant consequences of simulating few objects are the smoothing of possible transitions (e.g. percolation, phase separation) and an overall worse sampling of the quantities of interest, such as mean square displacements, structure factors, correlation functions, etc. To make things worse, the time scale for the diffusion of a single DNA construct scales with its mass, i.e. with the number of nucleotides it is made of. Since the characteristic time of the model is fixed and is linked to the motion of the single nucleotides, DNA constructs, which are usually made of tens to hundreds of nucleotides, diffuse very slowly. Therefore, the simulation of low-density systems requires very long times in order not to be diffusion-limited.

Notwithstanding these limitations, highly detailed simulations of all-DNA materials can be invaluable tools. Indeed, they allow for a close, even quantitative, comparison with experimental results [101]. Even more importantly, the high resolution of the model makes it possible to compute quantities which are not available in experiments, such as the conformation and flexibility of the constructs in the bulk, providing a strong link between the properties of the single constituents and the bulk behaviour [65].

For example, DNA trimers and tetramers, i.e. DNA constructs with valence 3 and 4, respectively, have been recently synthesised and investigated. The phase behaviour of these DNA nanostars at low salt concentration has been investigated experimentally, finding a gas-liquid-like phase separation at low temperature and concentration [19], as well as a rich and unusual dynamics [63, 64]. In order to gather a microscopic insight, nucleotide-level numerical simulations of 100 tetramers (19,400 nucleotides) were performed [65]. Unfortunately, for performance reasons the simulations had to be run at a higher salt concentration than experiment, preventing a direct, quantitative comparison with experiments. Thanks to lengthy

GPU-powered simulations, Rovigatti et al. have sketched the phase diagram of the system, evaluating the gas-liquid transition region, the percolation line and the isodiffusivity lines. These results demonstrate that the net effect of increasing the salt concentration of the solution is to increase the critical temperature and the width of the phase-separation region [102].

From a microscopic standpoint, the highly detailed description of DNA made it possible to probe the conformation of the single tetramers, and a quantitative match between neutron-scattering data and simulation results was found [101]. The analysis showed that tetramers retain their zero-density conformation in the bulk for all the investigated densities, an important property that further strengthens the connection between single-particle and bulk properties [65]. These results show that DNA tetramers are extremely flexible, and this flexibility is retained in the bulk, where they are part of a persistent network and form a gel-like material. It was later shown that such a high flexibility makes the low-temperature disordered phase the thermodynamic stable state, explaining why experiments did not find any sign of crystallisation. This result was obtained by performing rigorous free-energy calculations using simulations done with oxDNA [66]. From a microscopic standpoint, the average conformation of the tetramers, and the distance between bonded pairs and the angle formed by bonded triplets were all calculated, and their change with temperature and concentration evaluated. These quantities are important features of the building blocks and deeply affect the collective behaviour of the system. Therefore, their evaluation in realistic simulations makes it possible to design the strand sequences and tune the properties of the resulting materials in silico.

3.4.3   Indirect Simulations

Results from nucleotide-level simulations can be used as a starting point for an additional coarse-graining, exploiting these results to develop theoretical and numerical models, via either top-down or bottom-up strategies, that describe DNA constructs on mesoscopic length scales. These simpler models can then be used to probe the bulk phases with more ease. Here we will provide two examples: the investigation of the bulk behaviour of DNA nanostars and of the isotropic-nematic transition in DNA liquid crystals.

3.4.3.1 DNA Nanostars

As highlighted in the previous section, nucleotide-level simulations of bulk systems of DNA nanostars are feasible only for small systems, at high salt concentrations and moderate temperatures, and even these require very long simulation times. Here we explore two general strategies that can be used to overcome these limitations.

The first strategy we consider is purely numerical and consists in reducing the number of force sites required to simulate a DNA nanostar, and in particular a DNA tetramer, by performing a bottom-up coarse-graining. For example, the centre and arms of a tetramer can be well-represented by spheres connected by non-elastic

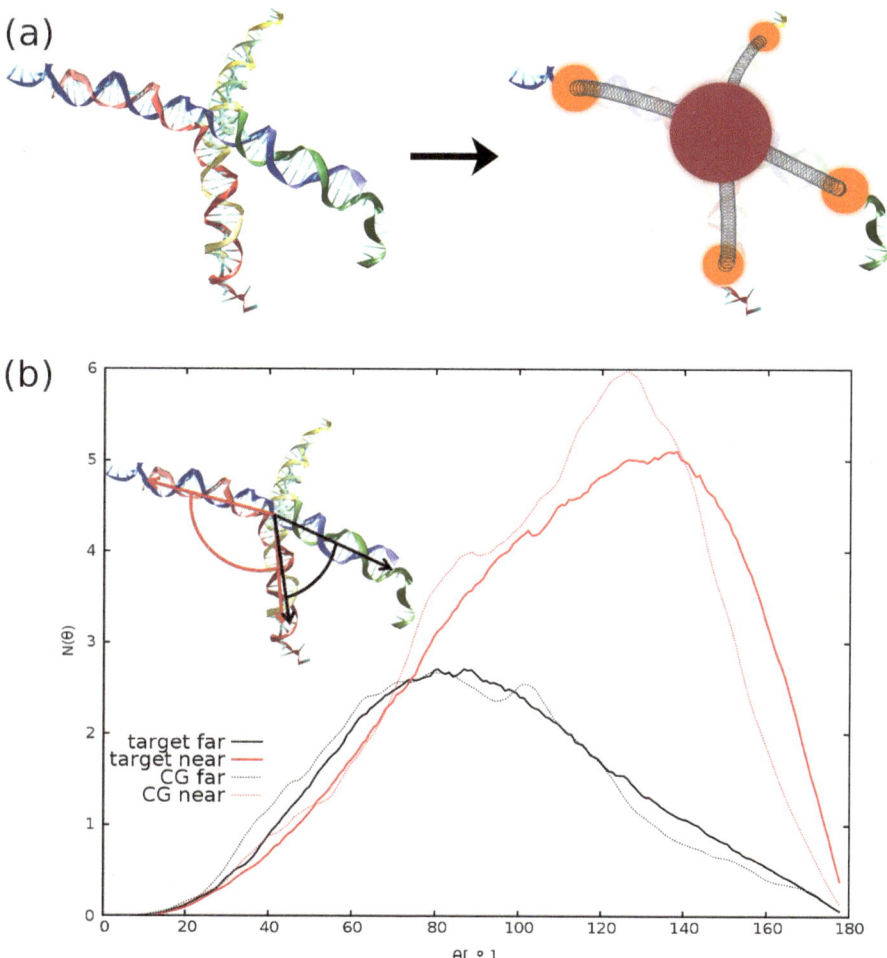

Fig. 3.2 (**a**) A DNA tetramer modelled (left) with oxDNA and (right) as five spheres connected by springs decorated with sticky ends. (**b**) Comparison between the distributions of the angles between near (red) and far (black) sticky ends, as indicated in the tetramer cartoon. Solid (dotted) lines refer to the full-detailed (coarse-grained) model. Figures adapted from [103]

springs, as shown in Fig. 3.2a. The effective interactions between these spheres can then be computed by employing an Iterative Boltzmann Inversion scheme, as done in [103]. Figure 3.2b presents the distributions of internal angles in the oxDNA and coarse-grained representations. These results show that the coarse-grained model captures the internal fluctuations of the tetramer conformations, as well as its overall shape and mechanics. The repulsion between two tetramers, being one of the quantities the model is fitted to, is also captured. The patchy-like connectivity can be reintroduced by connecting the single-stranded sticky ends, modelled through oxDNA, to the outer spheres, as outlined in Fig. 3.2a. Large-scale bulk simulations show that oxDNA and the coarse-grained model yield the same average number of bonds in the investigated concentration range. However, the radial distribution functions of the two systems turned out to be substantially different, owing to the excessive flexibility of the sticky ends in the coarse-grained model [103]. This issue, which calls for the addition of a multi-body potential to increase the stiffness of bonded arms in order to make it more accurate, highlights a general issue in coarse-graining procedures: some properties may be very well conserved by the coarse graining while some others may be not.

The second strategy we present makes use of fast two-body simulations to compute the input parameters required by the Wertheim thermodynamic perturbation theory (WTPT), which is a theoretical framework aimed at the evaluation of the thermodynamics of associating fluids [104, 105]. WTPT provides a closed expression for the free energy of the system as the sum of the free energy of a reference system and a term which takes into account the inter-particle bonding. For DNA systems, the latter can be readily evaluated through SantaLucia's nearest-neighbour model [74]. The former is usually approximated with the free energy of a system interacting through a purely repulsive potential. For colloidal systems the hard-sphere fluid, whose free energy is known with high accuracy, is often chosen as the reference system [106]. For DNA nanostars, whose mutual repulsion is soft in nature and is highly dependent on the size of the construct and on the salt concentration, a more appropriate choice is needed. In [102] it was shown that a virial expansion truncated at the first order, thus taking into account the repulsion due to pairs of nanostars only, results in a good agreement with experimental results. The evaluation of the two-body contribution, which requires the computation of the second virial coefficient of purely repulsive nanostars, can be very efficiently performed with oxDNA by employing a generalised Widom insertion method [107]. A new computation is required at each value of T and c, but thanks to WTPT this allows to compute instantaneously the total free energy of the system as a function of DNA concentration. Evaluation of the phase diagrams thus requires several calculations of a two-body quantity in the interesting range of temperature and ionic strength. Each of these calculations, depending on the parameters, takes from a few minutes to a few hours. Comparing these figures with the huge computer power and time required for the full simulations (see Sect. 3.4.2) demonstrates the benefit of employing this mixed theoretical/numerical approach to evaluate the phase diagram of these systems.

3.4.3.2 DNA Liquid Crystals

Concentrated solutions of DNA duplexes are known to exhibit liquid crystal phases [108, 109]. These states of matter are intermediate between the crystalline and liquid phases [110]. An example is provided by the nematic phase, in which particles display no positional order but are aligned along a common (nematic) axis. A classic argument by Onsager, derived in the framework of hard bodies, suggests that in order to exhibit a liquid crystal transition the basic building blocks of the system should possess a minimum degree of anisotropy [111]. Therefore, the appearance of liquid crystal phases in solutions of ultra-short (6–20 base pairs) dsDNA, with aspect ratio ranging from 1 to 4, cannot be explained by repulsive interactions alone. Indeed, blunt-ended duplexes experience a mutual attraction due to coaxial stacking interactions. As a net effect, duplexes stack onto each other and self-assemble into reversible chains that, provided that the concentration is high enough, align and form liquid crystal phases [57, 60, 112].

Such a system can be qualitatively investigated by using simple toy models such as cylinders with two patches, and the numerical results can be well-explained by a theoretical framework built on the theories developed by Wertheim and Onsager. However, this level of detail is not sufficient to establish a quantitative link with experiments. Unfortunately, the direct simulation of a DNA-based liquid crystal with a realistic model is currently too numerically demanding. To overcome this limitation De Michele and co-workers have shown that it is possible to employ quick simulations to estimate all the input parameters required by the theory to draw the whole phase diagram of the material [61]. These two inputs, namely the effective excluded volume and attractive interactions between two duplexes, can be evaluated by performing a Monte Carlo integral over the configurations of two duplexes modelled with oxDNA and require very little compute time. As a test, De Michele et al. used oxDNA to run bulk simulations showing that, in the limit of low density where these simulations are feasible, the theory is in quantitative agreement with the model. Considering that the two approaches yield a similar accuracy and that the large-scale simulations took 1–2 weeks on single GPUs (compared to a few hours on a single CPU core for the evaluation of the theoretical inputs), the theory developed in [61] allows to speed calculations up by orders of magnitude.

The main output of the parameter-free theoretical framework developed by De Michele et al. is the critical isotropic-nematic concentration, which depends on temperature and dsDNA length and which was shown to be in semi-quantitative agreement with experiments. As an additional outcome, these results suggested that the oxDNA model could benefit from a change in the coaxial stacking interaction. The proposed change has since been incorporated in the model [85]. A follow-up study demonstrated that, in order to increase the agreement with experiments, it is very important to take into account the precise geometric properties of the duplexes [113]. However, the strong dependence of these structural properties on the sequence can only be captured by all-atom simulations, providing an example of how it is often necessary to combine approaches of different level of detail to tackle the self-assembly behaviour of DNA-based systems.

3.5   Conclusions

The DNA molecule fulfils a very specific role in biology: it encodes the genetic information of all living beings. This is made possible by a very specific molecular recognition mechanism, the Watson–Crick pairing, and by the whole cell machinery that nature has developed in order to work with DNA. These two key features, together with the intrinsic compatibility of DNA with biological matter, make it possible to exploit DNA in countless applications, from drug delivery to materials science. The role of DNA in these applications can vary greatly, as it can act as the main component of the system [10, 19], as a component in a mixture [114], as a stabiliser [115], as a linker [21] or as a scaffold [14]. The characteristic length- and time-scales associated to these systems can differ by many orders of magnitude, making it impossible to use the same level of detail to carry out theoretical and computational investigations. As a consequence, recent years have seen a flourishing of new methods and models that describe DNA at different granularities. In this contribution we have first highlighted the challenges that need to be addressed in the fast-moving field of DNA-based materials. Then, we have shown that nucleotide-level coarse-grained DNA models can provide valuable insight into the processes which drive the self-assembly of these new-generation materials. Indeed, recent work has showed that this level of detail can be used to either directly probe the bulk phase behaviour or as a starting point for further coarse-graining, making these models invaluable assets for the development of the multi-scale approaches required to tackle the investigation of DNA-based systems.

Acknowledgements Lorenzo Rovigatti thanks Manfredo Di Porcia for providing the data reported in Fig. 3.2.

References

1. Egli M, Saenger W. Principles of nucleic acid structure. Berlin: Springer; 2013.
2. Watson JD, Crick FH, et al. Nature. 1953;171(4356), 737.
3. Hoogsteen K. Acta Crystallogr. 1959;12(10):822. https://doi.org/10.1107/S0365110X59002389. http://scripts.iucr.org/cgi-bin/paper?a02658.
4. Huppert JL. FEBS J. 2010;277(17):3452. https://doi.org/10.1111/j.1742-4658.2010.07758.x. http://www.ncbi.nlm.nih.gov/pubmed/20670279.
5. Seeman NC. J Theor Biol. 1982;99(2):237. https://doi.org/10.1016/0022-5193(82)90002-9.
6. Seeman NC. Annu Rev Biophys Biomol Struct. 1998;27(1):225. https://doi.org/10.1146/annurev.biophys.27.1.225.
7. Wollman AJM, Sanchez-Cano C, Carstairs HMJ, Cross RA, Turberfield AJ. Nat Nanotechnol. 2013;9:44.
8. Soloveichik D, Seelig G, Winfree E. Proc Natl Acad Sci. 2010;107:5393. https://doi.org/10.1073/pnas.0909380107.
9. He Y, Ye T, Su M, Zhang C, Ribbe AE, Jiang W, Mao C. Nature. 2008;452(7184):198.
10. Winfree E, Liu F, Wenzler LA, Seeman NC, Nature. 1998;394(6693):539.
11. Liu D, Park SH, Reif JH, LaBean TH. Proc Natl Acad Sci. 2004;101(3):717.
12. Yin P, Hariadi RF, Sahu S, Choi HM, Park SH, LaBean TH, Reif JH. Science. 2008;321(5890):824.
13. Rothemund PWK. Nature. 2006;440(7082):297.
14. Rajendran A, Endo M, Sugiyama H. Angew Chem Int Ed. 2012;51(4):874. https://doi.org/10.1002/anie.201102113.
15. Tsukanov R, Tomov TE, Masoud R, Drory H, Plavner N, Liber M, Nir E. J Phys Chem B. 2013;117(40):11932. https://doi.org/10.1021/jp4059214. PMID: 24041226.
16. Douglas SM, Bachelet I, Church GM. Science. 2012;335(6070):831. https://doi.org/10.1126/science.1214081.

17. Mirkin CA, Letsinger RL, Mucic RC, Storhoffand JJ, Nature. 1996;382:607. https://doi.org/10.1038/382607a0.
18. Seeman NC. Nature. 2003;421(6921):427.
19. Biffi S, Cerbino R, Bomboi F, Paraboschi EM, Asselta R, Sciortino F, Bellini T. Proc Natl Acad Sci. 2013;110(39):15633. https://doi.org/10.1073/pnas.1304632110.
20. Jones MR, Seeman NC, Mirkin CA, Science. 2015;347(6224):1260901. https://doi.org/10.1126/science.1260901. http://www.sciencemag.org/content/347/6224/1260901.
21. Maye MM, Kumara MT, Nykypanchuk D, Sherman WB, Gang O. Nat Nanotechnol. 2010;5(2):116.
22. Varrato F, Di Michele L, Belushkin M, Dorsaz N, Nathan SH, Eiser E, Foffi G. Proc Natl Acad Sci. 2012. https://doi.org/10.1073/pnas.1214971109. http://www.pnas.org/content/early/2012/10/31/1214971109.abstract.
23. Di Michele L, Eiser E. Phys Chem Chem Phys. 2013;15(9):3115. https://doi.org/10.1039/C3CP43841D.
24. Winfree E, Liu F, Wenzler LA, Seeman NC. Nature. 1998;394(6693):539.
25. Tkachenko AV. Phys Rev Lett. 2002;89(14):148303. https://doi.org/10.1103/PhysRevLett.89.148303. http://journals.aps.org/prl/abstract/10.1103/PhysRevLett.89.148303.
26. Lukatsky DB, Mulder BM, Frenkel D. J Phys Condens Matter. 2006;18(18):S567. https://doi.org/10.1088/0953-8984/18/18/S05. http://iopscience.iop.org/article/10.1088/0953-8984/18/18/S05.
27. Knorowski C, Burleigh S, Travesset A. Phys Rev Lett. 2011;106(21):215501. https://doi.org/10.1103/PhysRevLett.106.215501. http://link.aps.org/doi/10.1103/PhysRevLett.106.215501.
28. Angioletti-Uberti S, Mognetti BM, Frenkel D. Nat Mater. 2012;11(6):518.
29. Hagerman PJ. Annu Rev Biophys Chem. 1988;17(1):265.
30. Bloomfield VA. Biopolymers. 1997;44(3):269. https://doi.org/10.1002/(SICI)1097-0282(1997)44:3<269::AID-BIP6>3.0.CO;2-T.
31. Wood JL. Biochem J. 1974;143:775.
32. Lando DY, Haroutiunian SG, Kul'ba AM, Dalian EB, Orioli P, Mangani S, Akhrem AA. J Biomol Struct Dyn. 1994;12(2):355. https://doi.org/10.1080/07391102.1994.10508745. PMID: 7702774.
33. Williams MC, Wenner JR, Rouzina I, Bloomfield VA. Biophys J. 2001;80(2):874. https://doi.org/10.1016/S0006-3495(01)76066-3. http://www.sciencedirect.com/science/article/pii/S0006349501760663.
34. Tempestini A, Cassina V, Brogioli D, Ziano R, Erba S, Giovannoni R, Cerrito MG, Salerno D, Mantegazza F. Nucleic Acids Res. 2013;41(3):2009. https://doi.org/10.1093/nar/gks1206. http://nar.oxfordjournals.org/content/41/3/2009.abstract.
35. Seeman NC, Wang H, Yang X, Liu F, Mao C, Sun W, Wenzler L, Shen Z, Sha R, Yan H, Wong MH, Sa-Ardyen P, Liu B, Qiu H, Li X, Qi J, Du SM, Zhang Y, Mueller JE, Fu TJ, Wang Y, Chen J. Nanotechnology. 1998;9(3):257. https://doi.org/10.1088/0957-4484/9/3/018. http://iopscience.iop.org/article/10.1088/0957-4484/9/3/018.
36. Chen JH, Seeman NC. Nature. 1991;350(6319):631. https://doi.org/10.1038/350631a0. http://europepmc.org/abstract/med/2017259.
37. Park N, Um SH, Funabashi H, Xu J, Luo D. Nat Mater. 2009;8(5):432.
38. Amir Y, Ben-Ishay E, Levner D, Ittah S, Abu-Horowitz A, Bachelet I. Nat Nanotechnol. 2014;9(5):353.
39. Largo J, Starr FW, Sciortino F. Langmuir. 2007;23:5896.
40. Theodorakis P, Fytas N, Kahl G, Dellago C. Condens Matter Phys. 2015;18(2):22801.
41. Napper DH. Polymeric stabilization of colloidal dispersions. Vol. 3. New York: Academic; 1983.
42. Nykypanchuk D, Maye MM, Van Der Lelie D, Gang O. Nature. 2008;451(7178):549.
43. Feng L, Dreyfus R, Sha R, Seeman NC, Chaikin PM. Adv Mater. 2013;25(20):2779. https://doi.org/10.1002/adma.201204864.
44. Wang Y, Wang Y, Breed DR, Manoharan VN, Feng L, Hollingsworth AD, Weck M, Pine DJ. Nature. 2012;491(7422):51.

45. Lu F, Yager KG, Zhang Y, Xin H, Gang O. Nat Commun. 2015;6:6912. https://doi.org/10.1038/ncomms7912. http://www.nature.com/ncomms/2015/150423/ncomms7912/full/ncomms7912.html.
46. Kim JW, Kim JH, Deaton R. Angew Chem Int Ed. 2011;50(39):9185. https://doi.org/10.1002/anie.201102342.
47. Suzuki K, Hosokawa K, Maeda M. J Am Chem Soc. 2009;131(22):7518. https://doi.org/10.1021/ja9011386. PMID: 19445511.
48. Halverson JD, Tkachenko AV. Phys Rev E. 2013;87(6):062310.
49. van der Meulen SA, Leunissen ME. J Am Chem Soc. 2013;135(40):15129.
50. Angioletti-Uberti S, Varilly P, Mognetti BM, Frenkel D. Phys Rev Lett. 2014;113(12):128303. http://link.aps.org/doi/10.1103/PhysRevLett.113.128303
51. Rogers WB, Manoharan VN. Science. 2015;347(6222):639.
52. Romano F, Sciortino F. Phys Rev Lett. 2015;114(7):078104.
53. Bomboi F, Romano F, Leo M, Fernandez-Castanon J, Cerbino R, Bellini T, Bordi F, Filetici P, Sciortino F. Nat Commun. 2016;7:13191.
54. Bianchi E, Largo J, Tartaglia P, Zaccarelli E, Sciortino F. Phys Rev Lett. 2006;97(16):168301.
55. Romano F, Sanz E, Sciortino F. J Chem Phys. 2011;134(17):174502. https://doi.org/10.1063/1.3578182.
56. Romano F, Sciortino F. Nat Commun. 2012;3:975.
57. Nakata M, Zanchetta G, Chapman BD, Jones CD, Cross JO, Pindak R, Bellini T, Clark NA. Science. 2007;318:1276.
58. Zanchetta G, Bellini T, Nakata M, Clark NA. J Am Chem Soc. 2008;130(39):12864. https://doi.org/10.1021/ja804718c.
59. Zanchetta G, Nakata M, Buscaglia M, Clark NA, Bellini T. J Phys Condens Matter. 2008;20(49):494214.
60. De Michele C, Bellini T, Sciortino F. Macromolecules. 2012;45(2):1090. https://doi.org/10.1021/ma201962x.
61. De Michele C, Rovigatti L, Bellini T, Sciortino F. Soft Matter. 2012;8(32):8388.
62. Nguyen KT, Battisti A, Ancora D, Sciortino F, De Michele C. Soft Matter. 2015;11:2934. https://doi.org/10.1039/C4SM01571A.
63. Biffi S, Cerbino R, Nava G, Bomboi F, Sciortino F, Bellini T. Soft Matter. 2015;11:3132. https://doi.org/10.1039/C4SM02144D.
64. Bomboi F, Biffi S, Cerbino R, Bellini T, Bordi F, Sciortino F. Eur Phys J E. 2015;38(6):64. https://doi.org/10.1140/epje/i2015-15064-9.
65. Rovigatti L, Bomboi F, Sciortino F. J Chem Phys. 2014;140(15):154903. https://doi.org/10.1063/1.4870467. http://scitation.aip.org/content/aip/journal/jcp/140/15/10.1063/1.4870467.
66. Rovigatti L, Smallenburg F, Romano F, Sciortino F. ACS Nano. 2014;8(4):3567.
67. Smallenburg F, Sciortino F. Nat Phys. 2013;9(9):554.
68. Swigon D. Mathematics of DNA structure, function and interactions. New York: Springer; 2009. p. 293–320.
69. Peters JP, James Maher L. Q Rev Biophys. 2010;43(1):23. https://doi.org/10.1017/S0033583510000077. http://www.pubmedcentral.nih.gov/articlerender.fcgi?artid=4190679&tool=pmcentrez&rendertype=abstract.
70. Svozil D, Hobza P, Šponer J. J Phys Chem B. 2010;114(2):1191. https://doi.org/10.1021/jp910788e.
71. Mládek A, Krepl M, Svozil D, Cech P, Otyepka M, Banáš P, Zgarbová M, Jurečka P, Sponer J. Phys Chem Chem Phys. 2013;15(19):7295. https://doi.org/10.1039/c3cp44383c. http://pubs.rsc.org/en/content/articlehtml/2013/cp/c3cp44383c.
72. Laughton CA, Harris SA. Wiley Interdiscip Rev Comput Mol Sci. 2011;1(4):590. https://doi.org/10.1002/wcms.46. http://doi.wiley.com/10.1002/wcms.46.
73. Maffeo C, Yoo J, Comer J, Wells DB, Luan B, Aksimentiev A. J Phys Condens Matter. 2014;26(41):413101. https://doi.org/10.1088/0953-8984/26/41/413101. http://iopscience.iop.org/article/10.1088/0953-8984/26/41/413101.
74. SantaLucia J. Proc Natl Acad Sci. 1998;95(4):1460.

75. Poland D. J Chem Phys. 1966;45(5):1464. https://doi.org/10.1063/1.1727786. http://scitation.aip.org/content/aip/journal/jcp/45/5/10.1063/1.1727786.
76. Morriss-Andrews A, Rottler J, Plotkin SS. J Chem Phys. 2010;132(3):035105. https://doi.org/10.1063/1.3269994. http://scitation.aip.org/content/aip/journal/jcp/132/3/10.1063/1.3269994.
77. Savelyev A, Papoian GA. Proc Natl Acad Sci. 2010;107(47):20340. https://doi.org/10.1073/pnas.1001163107.
78. Cragnolini T, Derreumaux P, Pasquali S. J Phys Chem B. 2013;117(27):8047. https://doi.org/10.1021/jp400786b. PMID: 23730911.
79. Gonzalez O, Petkevičiūtė D, Maddocks JH. J Chem Phys. 2013;138(5):055102. https://doi.org/10.1063/1.4789411. http://scitation.aip.org/content/aip/journal/jcp/138/5/10.1063/1.4789411.
80. Linak MC, Tourdot R, Dorfman KD. J Chem Phys. 2011;135(20):205102. https://doi.org/10.1063/1.3662137. http://scitation.aip.org/content/aip/journal/jcp/135/20/10.1063/1.3662137.
81. Araque JC, Panagiotopoulos AZ, Robert MA. J Chem Phys. 2011;134(16):165103. https://doi.org/10.1063/1.3568145. http://scitation.aip.org/content/aip/journal/jcp/134/16/10.1063/1.3568145.
82. Hinckley DM, Freeman GS, Whitmer JK, de Pablo JJ. J Chem Phys. 2013;139(14):144903. https://doi.org/10.1063/1.4822042. http://scitation.aip.org/content/aip/journal/jcp/139/14/10.1063/1.4822042.
83. Hinckley DM, Lequieu JP, de Pablo JJ. J Chem Phys. 2014;141(3):035102. https://doi.org/10.1063/1.4886336. http://scitation.aip.org/content/aip/journal/jcp/141/3/10.1063/1.4886336.
84. Doye JP, Ouldridge TE, Louis AA, Romano F, Šulc P, Matek C, Snodin BE, Rovigatti L, Schreck JS, Harrison RM, et al. Phys Chem Chem Phys. 2013;15(47):20395.
85. Snodin BEK, Randisi F, Mosayebi M, Šulc P, Schreck JS, Romano F, Ouldridge TE, Tsukanov R, Nir E, Louis AA, Doye JPK. J Chem Phys. 2015;142(23):234901. https://doi.org/10.1063/1.4921957. http://scitation.aip.org/content/aip/journal/jcp/142/23/10.1063/1.4921957.
86. Hinckley DM, de Pablo JJ. J Chem Theory Comput. 2015;11(11):5436. https://doi.org/10.1021/acs.jctc.5b00341. PMID: 26574332.
87. Louis A. J Phys Condens Matter. 2002;14(40):9187.
88. Martinez-Veracoechea FJ, Mladek BM, Tkachenko AV, Frenkel D. Phys Rev Lett. 2011;107:045902. https://doi.org/10.1103/PhysRevLett.107.045902. http://link.aps.org/doi/10.1103/PhysRevLett.107.045902.
89. Romano F, Sciortino F. Phys Rev Lett. 2015;114:078104. https://doi.org/10.1103/PhysRevLett.114.078104. http://link.aps.org/doi/10.1103/PhysRevLett.114.078104.
90. Ouldridge TE, Louis AA, Doye JP. J Chem Phys. 2011;134(8):085101.
91. Ouldridge TE. Coarse-grained modelling of DNA and DNA self-assembly: coarse-grained modelling of DNA and DNA self-assembly. Berlin: Springer; 2012.
92. Šulc P, Romano F, Ouldridge TE, Rovigatti L, Doye JP, Louis AA. J Chem Phys. 2012;137(13):135101.
93. Šulc P, Romano F, Ouldridge TE, Doye JPK, Louis AA. J Chem Phys. 2014;140(23):235102.
94. Ouldridge TE, Šulc P, Romano F, Doye JP, Louis AA. Nucleic Acids Res. 2013;41(19):8886.
95. Schreck JS, Ouldridge TE, Romano F, Šulc P, Shaw LP, Louis AA, Doye JP. Nucleic Acids Res. 2015. https://doi.org/10.1093/nar/gkv582. http://nar.oxfordjournals.org/content/early/2015/06/08/nar.gkv582.abstract.
96. Matek C, Ouldridge TE, Doye JP, Louis AA. Sci Rep. 2015;5:7655.
97. Srinivas N, Ouldridge TE, Šulc P, Schaeffer JM, Yurke B, Louis AA, Doye JP, Winfree E. Nucleic Acids Res. 2013;41(22):10641.
98. Snodin BEK, Romano F, Rovigatti L, Ouldridge TE, Louis AA, Doye JPK. ACS Nano. 2016. https://doi.org/10.1021/acsnano.5b05865. http://pubs.acs.org/doi/10.1021/acsnano.5b05865.
99. Mladek BM, Fornleitner J, Martinez-Veracoechea FJ, Dawid A, Frenkel D. Phys Rev Lett. 2012;108(26):268301. https://doi.org/10.1103/PhysRevLett.108.268301.
100. Rovigatti L, Šulc P, Reguly IZ, Romano F. J Comput Chem. 2015;36:1. https://doi.org/10.1002/jcc.23763.

101. Fernandez-Castanon J, Bomboi F, Rovigatti L, Zanatta M, Paciaroni A, Comez L, Porcar L, Jafta CJ, Fadda GC, Bellini T, Sciortino F. J Chem Phys. 2016;145(8):84910. https://doi.org/10.1063/1.4961398.
102. Locatelli E, Handle PH, Likos CN, Sciortino F, Rovigatti L. ACS Nano. 2017;11(2):2094. https://doi.org/10.1021/acsnano.6b08287.
103. Di Porcia M. Coarse grained simulation of DNA tetramers. Master Thesis. Sapienza Università di Roma; 2015.
104. Wertheim MS. J Stat Phys. 1984;35:19. https://doi.org/10.1007/BF01017362.
105. Wertheim MS. J Stat Phys. 1984;35:35. https://doi.org/10.1007/BF01017363.
106. Fantoni R, Pastore G. Mol Phys. 2015;113(17–18):2593. https://doi.org/10.1080/00268976.2015.1061150.
107. Mladek BM, Frenkel D. Soft Matter. 2011;7:1450. https://doi.org/10.1039/C0SM00815J.
108. Strzelecka TE, Davidson MW, Rill RL. Nature. 1988;331(6155):457.
109. Bellini T, Cerbino R, Zanchetta G. In: Tschierske C, editor. Liquid crystals - materials design and self-assembly. Topics in current chemistry. Vol. 318. Berlin: Springer; 2012. p. 225–79. https://doi.org/10.1007/128_2011_230.
110. Prost J. The physics of liquid crystals. Vol. 83. Oxford: Oxford University Press; 1995.
111. Onsager L. Ann N Y Acad Sci. 1949;51(4):627. https://doi.org/10.1111/j.1749-6632.1949.tb27296.x. http://doi.wiley.com/10.1111/j.1749-6632.1949.tb27296.x.
112. Zanchetta G, Giavazzi F, Nakata M, Buscaglia M, Cerbino R, Clark NA, Bellini T. Proc Natl Acad Sci USA. 2010;107(41):17497. https://doi.org/10.1073/pnas.1011199107.
113. Nguyen KT, Battisti A, Ancora D, Sciortino F, De Michele C. Soft Matter. 2015;11(15):2934. https://doi.org/10.1039/c4sm01571a. http://pubs.rsc.org/en/Content/ArticleHTML/2015/SM/C4SM01571A.
114. Liu W, Tagawa M, Xin HL, Wang T, Emamy H, Li H, Yager KG, Starr FW, Tkachenko AV, Gang O. Science. 2016;351(6273):582. https://doi.org/10.1126/science.aad2080. http://science.sciencemag.org/content/351/6273/582.abstract.
115. Kegler K, Salomo M, Kremer F. Phys Rev Lett. 2007;98(5):058304. https://doi.org/10.1103/PhysRevLett.98.058304. http://journals.aps.org/prl/abstract/10.1103/PhysRevLett.98.058304.

Experimental Study of Self-Assembling Systems Characterized by Directional Interactions

4

Peter van Oostrum

4.1 Colloidal Self-Assembly

Self-assembly is essentially the formation of some kind of order, perceived by humans and possibly quantified by some order parameter, as a result of a competition between various enthalpic and entropic factors. This holds in atomic or (bio)molecular systems as well as on much larger length scales, as for instance in hierarchical assemblies or in colloidal systems, where the self-assembling units are typically three orders of magnitude larger than atoms and molecules. Colloidal particles used in experimental self-assembly studies are often a few micrometers in diameter. The great advantage of colloidal self-assembly over atomic self-assembly from a research perspective is that it is quite straightforward to track the individual particles during assembly using optical microscopy techniques [1].

In fact colloidal particles were observed as they assembled in a range of different phases such as gas, liquid [2], gels [3], glasses [4] and several crystal phases [5, 6]. The relatively large dimensions of colloidal particles allow for a complex architecture in that they can be equipped with a broad range of possible surface chemistries to modify the interaction potential between them. Moreover, the composition of the solvent can also be varied with strong effects on the effective particle–particle interaction potentials. The amount of material in the interior of colloidal particles makes them susceptible to external fields. These fields result in forces on the collective particles that can, for instance, be used to control their local concentration. A well-known example of how colloidal dispersions can be used to do research on self-assembly is in fact based on gravity. Gravity pulls particles down, lowering their potential energy, or enthalpy, and this competes with the

P. van Oostrum (✉)
Department of Nanobiotechnology, Institute for Biologically inspired materials, BOKU - University of Natural Resources and Life Sciences Vienna, Muthgasse 11-II, 1190 Vienna, Austria
e-mail: peter.van.oostrum@boku.ac.at

© Springer International Publishing AG, part of Springer Nature 2017
I. Coluzza (ed.), *Design of Self-Assembling Materials*,
https://doi.org/10.1007/978-3-319-71578-0_4

91

entropy that promotes the mixing of the particles throughout the sample volume. These two competing factors reach a thermodynamic equilibrium, the hydrostatic equilibrium, which for dilute dispersions, in which the particles are assumed to not interact amongst each other, is referred to as the barometric equilibrium. In the barometric equilibrium the number density η of the particles decays in an exponential manner, $\eta(z) \approx \exp(-z/L)$, with the height z; the decay constant $L = k_B T/mg$ is the gravitational length, with T the absolute temperature, m the buoyant mass of a single particle and g the gravitational height. The gravitational length of a dispersion of pollen was measured famously by Perrin to determine Boltzmann's constant, k_B [7]. Gravitation has been explicitly used to concentrate dispersions of spherical colloids in a controlled manner. The interactions between so-called "hard spheres", particles that repel each other like marbles as soon as they touch, lead to a concentration dependent phase behaviour that was mapped out using gravity to concentrate the particles [4]. Another example of the use of the large susceptibility of colloidal particles to external fields to locally concentrate them and study their phase behaviour is found in the use of the so-called "gradient force", the interaction between electric field induced dipoles in the particles with gradients in the exciting field strength [8–11]. Also here, there is a competition taking place between the enthalpy of the polarization state which concentrates the particles, in either a high field or a low field region of the sample depending on the sign of the polarizability at the applied frequency on the one hand, and the entropy which is maximum if the particles are homogeneously distributed throughout the sample, on the other hand.

Apart from the effects on the local concentration of colloidal particles, external fields can even give rise to interactions between individual particles [12, 13]. A nonexhaustive but concise overview of possible colloidal interactions is given in reference [14] which provides a good impression of what knobs can be turned with colloidal interactions.

The large amount of material colloidal particles consist of and their resulting large mass is a great help in the lab as it allows to very rapidly and efficiently separate the particles from the dispersing medium by centrifugation. On the other hand, the large mass and polarizability of colloidal particles also brings experimental challenges. The tendency of particles to sediment, or, less frequently cream, is an obstacle for the experimenter that wants to study self-assembly in the *bulk*, away from any walls or interfaces. Sedimentation or creaming can obviously be reduced by choosing to work with significantly smaller particles: often the gravitational length has to be large enough to make sedimentation irrelevant on the timescale of the experiment. However, for too small units, the visualization of individual particles is challenging, especially if the particles in the occurring phase are nearly touching. The used microscopy technique sets the lower bound to the particles size. To facilitate the microscopic study of any self-assembled structure, especially when the used microscopic contrast is based on fluorescence, it is advantageous that the particles are dispersed in a refractive index matching solvent; a solvent with the same refractive index as that of the particles [1]. Refractive index matching reduces the scattering of the light by particles above and below the specific particles in the field of view, along with the resulting noise. Like the sedimentation, the scattering of light by the particles can be reduced by using smaller particles. Again, the lower limit to the particles size is set by the resolution of the used microscopic technique. Last but not least, it is important to say that the speed of diffusion and therefore the rate at which any self-assembly takes place depends strongly on the size of the particles, and on the viscosity of the medium. The composition of a colloidal system for self-assembly experiments therefore is a compromise between requirements on the time scale, the length scale, the density, the refractive index and all those properties that determine the colloidal interactions that should ideally lead to the desired assembly behaviour.

The example of hard sphere crystallization we just touched upon has been very important in the development of a thorough understanding of the key role played by entropy in self-assembly processes. The fact that hard spheres only repel each other means that they cannot release heat upon binding to compensate for the decrease of entropy associated with the formation of crystalline order. What happens is that when the particles are confined to a regular lattice they all have more space to rattle around their equilibrium positions than they would have had in a more random or glass-like arrangement and thus crystallization is driven by the vibrational entropy (at the cost of the translational entropy). There are many examples in which entropy plays an important role, many of which are colloidal in scale. A very nice commentary on the role of entropy in self-assembly is given here [15].

4.2   Anisotropically Interacting Particles

Despite the extensive control experimentalists have over the interactions, colloidal self-assembly of isotropically interacting particles cannot be fully commanded. That is to say: one cannot direct the self-assembly between simple colloids towards any conceivable structure. In order to gain a greater control over the self-assembly, one would like to impose further constraints on the systems. Extra *instructions* for self-assembly can be imparted upon the particles if the interactions are no longer merely isotropic but rather depend on the relative positions and orientations of the particles. Often such anisotropic interactions can be realized by selectively modifying part of the surface of the particles. The different surface areas on the colloids are referred to as *patches* and during the last few decades many different techniques to synthesize or modify patchy particles have been published [16–20]. Other kinds of anisotropic interactions can arise during particle synthesis. For example, field induced electric or magnetic dipolar interactions are a middle way between isotropic and patchy interactions in that the orientation of the anisotropic potential around a particle does not reorient with the particle but rather with the external field. Finally, the interactions between anisotropically shaped particles is inherently anisotropic, mainly through entropic effects.

For the study of self-assembly in colloidal systems in which anisotropy plays an important role, the same basic criteria apply as for isotropically interacting systems: the particles should be large enough to distinguish them at the shortest possible distance between them with the microscopy technique used and, in case the assembly in the bulk is object of study, the gravitational length must be such that sedimentation is irrelevant on the timescale of the experiment. Often it is also desired that the orientational order of anisotropic particles can be probed. That essentially means that sub-particle details must be distinguishable, which in turn implies that one needs a better microscope or larger particles. To achieve the best resolution a microscope can offer, it is important that undesired scattering is minimized; this is realized when the refractive index of the particles is as close as possible to that of the dispersing medium. Clearly for more complex particles with an internal structure this is an extra constraint for the choices of material.

We will first shortly revisit a few methods to realize anisotropic interactions between colloidal particles that all have in common that the patch size and distribution is determined during the synthesis while these properties remain unaltered during the consecutive self-assembly. These methods can be referred to as *top-down*: all the information for self-assembly is transferred to the particles during synthesis while any fluctuations are *frozen in* as permanent polydispersity. Of course, at a later stage, during the assembly the interactions can be modified by applying external fields and by changing the solvent composition in terms of salt, pH and possibly depletant concentration. We will then proceed with a discussion of a few examples in which anisotropically interacting units arise through self-assembly of smaller, possibly flexibly linked, subunits. This mechanism is sometimes called *bottom-up*. The advantage of the bottom-up approach is that the self-assembling units in principle are formed or rearranged also during the self-assembly of any larger structures. This is referred to as *hierarchical* self-assembly [21, 22].

We note that self-assembly is essentially a process in which some order arises spontaneously and thus it is difficult to draw a hard line between protocols to make patchy particles that make use of self-assembly, bottom-up, and protocols that do not, top-down. Here we make this distinction based on the answer to the question whether there are subsequent modification steps required after the formation of the anisotropically interacting units. The mechanism to the emergence of anisotropic interaction, we observe, can in itself arguably be called self-assembly in the vast majority of cases. Finally, we present some ideas about how more detailed and rephrasable instructions for directed self-assembly might be given to colloidal self-assembling systems.

4.2.1 Patchy Particles from the Top-Down Route

A very helpful classification of synthesis methods of patchy particles was presented in reference [16], although sometimes these strategies are combined. Other useful review papers on patchy colloids are [18] and [19]. Here we shortly revisit a selection of synthesis methods to provide an insight in their aptness to produce useful experimental model systems for the study of anisotropy driven self-assembly.

A common method for the fabrication of patches is based on the evaporative deposition of material on part of the surface of colloidal particles that are dried on a substrate. In the simplest case this yields particles with half of their surface covered with a metal or metal-oxide. Such particles are referred to as *Janus* particles after the Roman god with two faces [23]. The interested reader is referred to two excellent review papers specifically on Janus particles and the various methods in which they have been synthesized [24, 25]. More control over patches from evaporative deposition in terms of size, shape and number of patches can be obtained by *glancing angle deposition* [26] of, for instance, a gold layer on particles in a two-dimensional array followed by a thiol functionalization of the patches to provide the properties that lead to the desired interaction. This can, for instance, be used to make attractive, hydrophobic patches [27] or to make charged regions [28]. Glancing angle deposition techniques offer rather extensive possibilities to control the relative orientation and size of the patches; one can combine the use of different angles of deposition, the casting of shadows in ordered particle arrays and controlled etching of deposited patches to make these smaller with the possibility to lift off and flip the entire particle array to proceed with the modification of patches on the other side of the particles [26, 29–31]. Glancing angle deposition can be applied to make patches both on oxide particles like silica and on polymeric particles; the latter can relatively easily be density matched for self-assembly studies in the bulk. Since glacing angle deposition is a technique based on 2D arrays of particles, the amount of particles that can be modified in a single batch is limited. Moreover, when glancing angle deposition uses the casting of shadows of particles in regular arrays these indeed need to be very regular to yield monodisperse patches between different particles. It is worth mentioning that the deposited material forming the patch might interfere with light microscopy efforts to visualize any self-assembled structures in 3D due to the absorption of light.

Other methods are in one way or another based on *templating* where some sacrificial material is used to temporarily protect part of the particles' surface. Silica particles can be made to sink into mixed polymer beds to leave part of the particle surface free to surface modification. The affinity between the particles and the polymers can be chemical [32] or physical [33], while the depth of penetration and thus the patch size can be determined through the reaction time and the temperature. The actual formation of the patches is the result of a surface modification (often silanization) of the part of the particle surface that is not protected by the template. Usually these techniques yield particles with one patch per particle, i.e., Janus particles, but there are exceptions in which templating yields two patches on spherical particles [33, 34] or even non-spherical particles with two polymeric caps resulting from partial etching of electrospun polymeric wires containing particles [35]. The chemical anisotropy experienced at the surface of a colloidal particle can be brought about by placing the particles at a liquid–liquid [36–40] or liquid–gas [41, 42] interface where the contact angle, which in turn can be influenced with a surfactant, can be used to determine the position of the particle relative to the interface and thus the *Janus fraction*. A combination of hydrophobic and hydrophilic ligands has been used to realize an asymmetric surface chemistry on nanoparticles at the interface between oil and water [43, 44]. When emulsions are used, the yield can be relatively high with respect to when planar templates are used as this starts to approach bulk scaling.

Liquid interfaces and surface tension have been used to provide anisotropy in surprising ways. Particles on the surface of an emulsion droplet form regular aggregates upon evaporation of the solvent [45], which can be used to make patchy particles with different numbers of patches in regular distributions that can be separated from each other by density gradient centrifugation. Controlled patch sizes result upon swelling these regular aggregates with a liquid monomer to be later polymerized [46]. Interactions between the patches can be modified and made specific by functionalizing the patch surface with DNA [46] or metal-coordination-based recognition units [47]. The larger patches made in this way are protrusions on the particle surface. Alternative methods to make particles with protruding patches using surface tension are based on swelling a polymeric particle with additional monomer to form a controlled number of protrusions [48] or to condense monomer droplets onto an oxide particle [49], after which the monomers can be polymerized to make these patches permanent. This central oxide particle can be grown further and then the polymeric protrusions can be dissolved leaving a solid particle with a regular pattern of dimples [50]. Protrusions of monomer can also be polymerized to form areas with an effective surface roughness. In combination with depletion attractions induced by so-called depletants on the scale of the roughness on these patches this gives rise to neutral patches or protrusions on otherwise attractive particles [51]. Similarly, dimples left after a two-stage polymerization [52] or flat faces left after temporary melting at a flat substrate can be used to render inherently isotropic interactions such as depletion [51,53] and van der Waals attractions [54,55] directional.

Finally, a conceptually simple manner to make particles with one or two polar patches is based on contact printing, usually with a soft polydimethylsiloxane (PDMS) stamp on 2D particle arrays [56–58]. Similar stamps are in fact often used to lift off and flip the particle arrays for glancing angle deposition. Via the stiffness of the stamp, which can be easily modified through the mixing ratio between monomers and cross-linkers, the applied pressure and the amount of "ink" used, the size of the produced patches can be varied. Contact printing can also be applied to make patches on polymeric particles that can be density matched for self-assembly studies in three dimensions.

The spontaneous assembly of patchy colloids has been experimentally observed for instance, in the site-specific aggregation of finite colloidal clusters [46] or in the formation of an extended, two-dimensional crystal, known as kagome lattice [27]. The assembly of patchy particles can be further influenced with external fields, especially in the case of conducting patches on dielectric particles [59–63].

So far we have discussed how to create directional interactions between essentially spherical colloidal particles by creating differently interacting patches. There are also other methods to introduce anisotropy in the interactions. Shape anisotropy can be realized, for instance, by making use of different growth potentials of crystal planes [64], by embedding polymeric particles in a sacrificial polymer matrix that is then deformed [65] or even by swelling polymeric particles in a 3D colloidal crystal [66]. Non-spherical particles display a variety of interesting super-structures, such as cubic monolayers, fluid-like membranes that are reminiscent of lipid bilayers and monolayers of octapods [64, 67, 68]. The great potential offered by anisotropic particles is further extended if the anisotropy forms a handle to external fields. Under the influence of electric or magnetic fields, the formation of chains, staggered chains or close and loosely packed two-dimensional crystals can be induced [69]. An overview of experiments in which external fields under different orientations are used is given in reference [70]. By exposing multi-component dispersions to external fields and/or dispersing them in nematic solvents, more exotic structures with triangular-packed, square-packed and honeycomb arrays can be assembled [71–73]. Additionally, light [11, 74, 75] and magnetic [76] fields that exhibit intensity variations on the length scale of individual colloids can be used to shape the energy landscape, thus steering the assembly into (a)periodic or quasi-crystalline structures.

Most experimental work on patchy particles deals with systems of mutually repulsive particles with some attractive patches. However, there is a large class of systems with a heterogeneous surface charge. The interactions between charged particles with a small number of oppositely charged patches are rather complex as the patch–patch interactions are repulsive; the patches on such particles are attractive to the non-patch areas of other particles only. To highlight the contrast with respect to the more conventional attractive patches, particles with a patchy, heterogeneous, surface charge have been called *inverse patchy particles* [77]. The self-assembly of inverse patchy particles has been demonstrated to yield structures that have not been reported for conventional patchy particles. With relatively simple surface patterns [34] it is possible to steer the assembly into specific morphologies while the interactions can be tuned through two different and independent aspects of the dispersing medium: the pH and the screening length. For a more complete description of the modeling and self-assembly of IPCs, we refer to a recent review paper [78] and to Chap. 2, which gives concise summary of the coarse grained description of inverse patchy particles.

On a final note, self-assembly on occasion does play a role in the experimental protocols using the so-called top down approach. The formation of 2D crystals, for instance, by capillary forces during the drying of the dispersing medium [79], used for glancing angle patch depositions [26] or the ordering in templates [34] as well as the formation of a 3D colloidal crystal for particle deformation [66] and the formation of regular aggregates upon evaporation of emulsion droplets [45] are in itself self-assembly. As mentioned before, here we make the distinction between the top-down and bottom-up approach based on whether there are or not subsequent modification steps after the formation of the anisotropically interacting units.

4.2.2   Patchy Particles from the Bottom-Up Route

Conceptually the smallest step from particles of some solid material with one or more patches to self-assembled patchy units is probably taken in the direction of polymer brushes on nanoparticles that can respond to changes in the solvent quality by forming local aggregates that could be considered patches. The number of these patches depends on the lengths of the grafted polymers with respect to the particle diameter and the solvent quality. The patches can be rendered permanent by light induced cross-linking [80]. The idea behind this type of polymer based patchy particle is comparable with the formation of patches in so-called telechelic star polymers, star polymers with diblock copolymer arms [81]. Here the patches are even more mobile by virtue of the brush like nature of the cores. Possibly the patchy nature of telechelic star polymers in solution has been observed to give rise to self-assembly into large crystals [82, 83].

Surface tension between different liquids or polymer (mixtures) has been mentioned above as a driving force for the generation of some patchy or anisotropic particles. Below we add more examples of protocols that generate anisotropic particles or aggregates that all have in common that the anisotropy can in principle (be made to) rearrange.

By carefully choosing the monomers, polymers dissolved in monomer droplets in an emulsion undergo a micro phase separation due to a changing interfacial tension upon polymerization of the monomers. This leads to the formation of particles with attractive patches [84]. By rapidly injecting polymer blends in solution into a non-solvent it is anticipated that particles that are then forming [85] become patchy with some control over the distribution of the patches [86, 87]. Upon the slow evaporation of a liquid that is a good solvent to two different polymers from a mixture containing also a non-solvent to both polymers, Janus particles with different Janus fractions form [88]. The precise distribution of the two polymers throughout the resulting particles is controlled by the mixing ratios. If additional constraints are added, such as the links between some of these two polymers in case they are presented as block copolymer, various types of patchy and stripy particles can form [88]. In principle one can redissolve the polymers by adding more of the good solvent to change the composition of the particles, although the removal of the constraints formed by the block copolymer is non-trivial.

A nice example of the formation of anisotropic units by changing external parameters can be observed upon the mixing of two types of diblock copolymers that then can form micelles with the common polymer in their "core" and the two types of copolymer in their shell. The polymers in the shell can be driven to demix to transform these micelles into multi patch particles at low ionic strength or phase separate and break up into Janus particles upon a further reduction of the solvent quality through the pH [89]. Similar results can be obtained with triblock copolymers, where the multi-patch micelles are different in that the third polymer is "grafted" on the patches rather than in between the patches [90]. The patch forming polymers can be cross-linked to make permanent Janus particles. Changing the way in which three types of polymers are linked together into *hetero-* or *mikto*-arm star terpolymers changes the constraints on the rearrangements of the molecules which gives rise to yet different self-assembly scenarios: by changing the solubility of one of the arms, via the concentration of particular ions, it is possible to manipulate a micelle-like phase via cylindrical structures and superstructures thereof into a lamellar phase [91]. The same approach was used with linear triblock terpolymers that can be manipulated into forming patchy particles with two or three patches depending on the volumes occupied by the different blocks, a parameter that cannot be modified after the synthesis of the terpolymers: by modifying the solvent quality, it is possible to induce hierarchical self-assembly and co-assembly of different types of soft patchy particles that are self-assembly products themselves [92]. This allowed the researchers to reach an amazing level of control over the formed superstructures where size- and number-ratios determine the size distribution of the resulting *colloidal polymers* and larger self-assembled structures these in turn form on a substrate [92]. Although the formation of the patchy particles themselves is in effect *frozen in* by the insolubility of the cores of these self-assembled particles, in principle, they can be allowed to rearrange by changing the solvent conditions back.

It is worth mentioning that the real time and real space visualization of all these polymer-based systems through light microscopy is practically impossible because of the limited sizes of the forming particles and patches. These are in turn limited by the lengths of the polymers. Electron microscopy after drying the sample does allow to get an insight in the structures present in the sample, but the dynamics are hard to follow and one should be always ware of possible artifacts introduced by the sample preparation. The assembly dynamics can be probed on a collective, statistically perhaps more relevant, level through dynamic light scattering or small angle X-ray scattering as, for example, was used in reference [91].

4.3   Outlook: Colloidal Strings and Patchy Polymers

All the effort spent on the synthesis of patchy particles via top-down routes, or frozen bottom-up routes has led to the impressive realization of a few examples of self-assembly where the symmetry breaking at the single particle level allowed the design of somewhat more complex regular or amorphous structures [27, 46, 92]. Nevertheless the resulting materials remain relatively simple and systems of large addressable complexity will have to assemble at various levels and timescales [15].

To steer self-assembly into very complex and detailed designable materials one could use very specific interactions that drive particles to bond only to their desired neighbours, as for instance has been demonstrated with so-called DNA-bricks [93]. This is the "puzzle" approach to self-assembly [94]: it is possible to consider any structure to be self-assembled as a 3D picture, the pixels of which can be switched on, i.e., added to the mixture in which self-assembly is desired. The synthesis of many different particles with specific surface interactions, for instance through DNA, is demonstrably possible, but likely very costly. For more on self-assembly in DNA-based systems, we refer to Chap. 3.

An alternative approach to obtain highly specific self-assembly is to convey the information for the structure to be self-assembled to the system in the form of the sequence of simpler units, that can be recycled by concatenation in many alternative sequences. This is called the folding approach [94] and it is the strategy that is adopted by nature to make a vast range of different structures through, e.g., protein folding. The folding of proteins is perhaps the most complex, robust and abundant example of self-organization. By creating different peptide sequences of only 20 types of amino acids that self-fold into a broad variety of structures, nature uses self-organization to synthesize most of the substances needed to realize all known life forms [95, 96]. The underlying physics is much simpler than the many, partially quantum mechanical, aspects of the interactions between the amino acids and the surrounding molecules may lead us to believe [97–100]. Simulations have shown that control over the folding of chains can be obtained in model systems through the sequence of as little as two dissimilar units, a property that is called *designability* [101]. The instructions for the folding are in the sequence of simple units that can in principle be re-used to create many dissimilar concatenations.

The key ingredients for this utterly versatile type of self-organization are (1) the availability of chains of differently interacting units and (2) the control over the sequence of concatenation. A third ingredient is however necessary: merely a sequence of dissimilar isotropic interactions can not guarantee an accurate self-organization and therefore the configurational space needs to be further confined by the possible formation of (3) directional attractive bonds between the chain units. While in biopolymers such as proteins this confinement is realized by the hydrogen bonds between the amino acids [98], at the colloidal level, directional interactions are referred to as patchy interactions; hence the name *patchy polymers*. The described requirements are stringent but a wide range of different systems with interactions of various origin can in principle meet these conditions. The systems should be synthesized in such a way that the bonding energy of the patches is comparable in strength to the isotropic interactions and that these energies can compete with entropy. The simplicity of the model allows to explore sufficient sequences and configurations thereof to be able to make educated choices of structures that can be expected to be designable with a given set of interactions. A chosen specific shape can then be used as a target structure and various sequences with a high probability to fold into this configuration can be proposed. Consequently chains with a designed sequence of interacting units were shown to fold into the designed target structure [101]. The generality and versatility of this strategy to direct self-assembly has been demonstrated by the successful design and folding into target structures in systems of chains of 20, or only two different interacting potentials that were further confined with two or even only one directional interaction [101, 102]. Interestingly it has been demonstrated that even knot-like structures can be chosen as target for the folding and that in analogy with proteins such knotted configurations can significantly change, for instance, the melting temperature of these targets upon cyclization [102].

It has been possible to link colloidal particles, aligned in an external field, into stiff, semi-flexible or flexible strings with dynamics that are similar to those of molecular polymers. The particles can be linked by entanglement of polymers on the surface of the particles [103, 104], bifunctional molecules [105] or DNA linkers [106]. Some control over the inner order in chains of colloidal particles has been achieved via the alignment, using the torque in a direct current field, of heterogeneously charged PMMA-PS dumbbells and subsequent polarization with an alternating current field before linking the particles by entanglement [104]. Superparamagnetic polystyrene particles, i.e., polystyrene particles loaded with superparamagnetic iron oxide nanoparticles, can be magnetically polarized as well as, more generally, particles with an electric polarizability can be electrically polarized, to induce a direction dependent dipolar attraction that can overcome the charge stabilization to form strings. By varying the speed at which the polarizing fields are switched on, some control over the sequence of larger, more polarizable, and smaller particles in the strings can be obtained. After the particles assembled into strings they can be bonded permanently by heat fusion [13,107]. If the magnetic field is increased rapidly in bidisperse batches, random sequences of smaller and larger particles form, while if the field is switched on gradually, the larger particles form strings first, followed by the smaller particles giving rise to short blocks of particles with a similar size [108]. By mixing strings from different batches of particles colloidal analogues of block copolymers were realized [104]. In order to determine the sequence of different particle types along the chains one could develop a colloidal analogue of Solid Phase Peptide Synthesis [109]: by combining field induced dipoles to drive string formation in a simple microfluidics channel, it might be possible to stepwise add micrometric particles and chemical compounds to enact the bonding of single particles to the strings and modify their surface chemistry. This strategy to add very specific constraints at will offers a tantalizing perspective on the self-assembly of meta-materials from colloidal particles. Creating purely artificial systems with specific control over the folding would open new possibilities for the design of novel materials.

References

1. Van Blaaderen A, Imhof A, Hage W, Vrij A. Three-dimensional imaging of submicrometer colloidal particles in concentrated suspensions using confocal scanning laser microscopy. Langmuir. 1992;8(6):1514–7.
2. Aarts DGAL. Direct visual observation of thermal capillary waves. Science. 2004;304(5672):847–50.
3. Zaccarelli E. Colloidal gels: equilibrium and non-equilibrium routes. J Phys Condens Matter. 2007;19(32):323101.
4. Pusey PN, van Megen W. Phase behaviour of concentrated suspensions of nearly hard colloidal spheres. Nature. 1986;320(6060):340–2.
5. Yethiraj A, van Blaaderen A. A colloidal model system with an interaction tunable from hard sphere to soft and dipolar. Nature. 2003;421:513–7.
6. Palberg T. Crystallization kinetics of colloidal model suspensions: recent achievements and new perspectives. J Phys Condens Matter. 2014;26(33):333101.
7. Perrin J. Atoms. London: Constable; 1916.

8. Sullivan MT, Zhao K, Hollingsworth AD, Austin RH, Russel WB, Chaikin PM. An electric bottle for colloids. Phys Rev Lett. 2006;96(1):015703.

9. Leunissen ME, Sullivan MT, Chaikin PM, van Blaaderen A. Concentrating colloids with electric field gradients. I. Particle transport and growth mechanism of hard-sphere-like crystals in an electric bottle. J Chem Phys. 2008;128(16):164508.

10. Leunissen ME, van Blaaderen A. Concentrating colloids with electric field gradients. II. Phase transitions and crystal buckling of long-ranged repulsive charged spheres in an electric bottle. J Chem Phys. 2008;128(16):164509.

11. Hermes M, Vermolen ECM, Leunissen ME, Vossen DLJ, van Oostrum PDJ, Dijkstra M, van Blaaderen A. Nucleation of colloidal crystals on configurable seed structures. Soft Matter. 2011;7(10):4623.

12. Hynninen A-P, Dijkstra M. Phase diagram of dipolar hard and soft spheres: manipulation of colloidal crystal structures by an external field. Phys Rev Lett. 2005;94(13).

13. Smallenburg F, Vutukuri HR, Imhof A, van Blaaderen A, Dijkstra M. Self-assembly of colloidal particles into strings in a homogeneous external electric or magnetic field. J Phys Condens Matter. 2012;24(46):464113.

14. Yethiraj A. Tunable colloids: control of colloidal phase transitions with tunable interactions. Soft Matter. 2007;3(9):1099.

15. Frenkel D. Order through disorder. Nat Mater. 2015;14:9–12.

16. Pawar AB, Kretzschmar I. Fabrication, assembly, and application of patchy particles. Macromol Rapid Commun. 2010;31:150.

17. Bianchi E, Blaak R, Likos CN. Patchy colloids: state of the art and perspectives. Phys Chem Chem Phys. 2011;13:6397–410.

18. Yi G-R, Pine DJ, Sacanna S. Recent progress on patchy colloids and their self-assembly. J Phys Condens Matter. 2013;25(19):193101.

19. Duguet É, Hubert C, Chomette C, Perro A, Ravaine S. Patchy colloidal particles for programmed self-assembly. C R Chim. 2016;19(1–2):173–82.

20. Bianchi E, Capone B, Coluzza I, Rovigatti L, van Oostrum PDJ. Limiting the valence: advancements and new perspectives on patchy colloids, soft functionalized nanoparticles and biomolecules. Phys. Chem. Chem. Phys. 2017;19:19847–19868.

21. Rest C, Kandanelli R, Fernández G. Strategies to create hierarchical self-assembled structures via cooperative non-covalent interactions. Chem Soc Rev. 2015;44(8):2543–72.

22. Gârlea IC, Bianchi E, Capone B, Rovigatti L, Likos CN. Hierarchical self-organization of soft patchy nanoparticles into morphologically diverse aggregates. Curr Opin Colloid Interface Sci. 2017;30:1–7.

23. Casagrande C, Fabre P, Raphaël E, Veyssié M. "Janus Beads": realization and behaviour at water/oil interfaces. EPL. 1989;9(3):251.

24. Lattuada M, Alan Hatton T. Synthesis, properties and applications of Janus nanoparticles. Nano Today. 2011;6(3):286–308.

25. Walther A, Müller AHE. Janus particles: synthesis, self-assembly, physical properties, and applications. Chem Rev. 2013;113:5194–261.

26. Pawar AB, Kretzschmar I. Multifunctional patchy particles by glancing angle deposition. Langmuir. 2009;25(16):9057–63.

27. Chen Q, Bae SC, Granick S. Directed self-assembly of a colloidal kagome lattice. Nature. 2011;469:381–4.

28. Hong L, Cacciuto A, Luijten E, Granick S. Clusters of charged janus spheres. Nano Lett. 2006;6(11):2510–4.

29. Chen Q, Diesel E, Whitmer JK, Bae SC, Luijten E, Granick S. Triblock colloids for directed self-assembly. J Am Chem Soc. 2011;133(20):7725–7.

30. Pawar AB, Kretzschmar I. Patchy particles by glancing angle deposition. Langmuir. 2008;24:355–8.

31. He Z, Kretzschmar I. Template-assisted fabrication of patchy particles with uniform patches. Langmuir. 2012;28(26):9915–9.

32. McConnell MD, Kraeutler MJ, Yang S, Composto RJ. Patchy and multiregion janus particles with tunable optical properties. Nano Lett. 2010;10(2):603–9.
33. Lin C-C, Liao C-W, Chao Y-C, Kuo C. Fabrication and characterization of asymmetric janus and ternary particles. ACS Appl Mater Interfaces. 2010;2(11):3185–91.
34. van Oostrum PDJ, Hejazifar M, Niedermayer C, Reimhult E. Simple method for the synthesis of inverse patchy colloids. J Phys Condens Matter. 2015;27(23):234105.
35. Ding T, Tian Y, Liang K, Clays K, Song K, Yang G, Tung C-H. Anisotropic oxygen plasma etching of colloidal particles in electrospun fibers. Chem Commun (Camb). 2011;47(8):2429–31.
36. Takahara YK, Ikeda S, Ishino S, Tachi K, Ikeue K, Sakata T, Hasegawa T, Mori H, Matsumura M, Ohtani B. Asymmetrically modified silica particles: a simple particulate surfactant for stabilization of oil droplets in water. J Am Chem Soc. 2005;127(17):6271–5.
37. Hong L, Jiang S, Granick S. Simple method to produce janus colloidal particles in large quantity. Langmuir. 2006;22(23):9495–9.
38. Böker A, He J, Emrick T, Russell TP. Self-assembly of nanoparticles at interfaces. Soft Matter. 2007;3(10):1231.
39. Jiang S, Granick S. Controlling the geometry (Janus balance) of amphiphilic colloidal particles. Langmuir. 2008;24(6):2438–45.
40. Perro A, Meunier F, Schmitt V, Ravaine S. Production of large quantities of "Janus" nanoparticles using wax-in-water emulsions. Colloids Surf A Physicochem Eng Asp. 2009;332(1):57–62.
41. Petit L, Manaud JP, Mingotaud C, Ravaine S, Duguet E. Sub-micrometer silica spheres dissymmetrically decorated with gold nanoclusters. Mater Lett. 2001;51(6):478–84.
42. Sabapathy M, Shelke Y, Basavaraj MG, Mani E. Synthesis of non-spherical patchy particles at fluid–fluid interfaces via differential deformation and their self-assembly. Soft Matter. 2016;12(27):5950–8.
43. Vilain C, Goettmann F, Moores A, Le Floch P, Sanchez C. Study of metal nanoparticles stabilised by mixed ligand shell: a striking blue shift of the surface-plasmon band evidencing the formation of Janus nanoparticles. J Mater Chem. 2007;17:3509–14.
44. Andala DM, Shin SHR, Lee H-Y, Bishop KJM. Templated synthesis of amphiphilic nanoparticles at the liquid–liquid interface. ACS Nano. 2012;6(2):1044–50.
45. Manoharan VN. Dense packing and symmetry in small clusters of microspheres. Science. 2003;301(5632):483–7.
46. Wang Y, Wang Y, Breed DR, Manoharan VN, Feng L, Hollingsworth AD, Weck M, Pine DJ. Colloids with valence and specific directional bonding. Nature. 2012;491(7422):51–5.
47. Wang Y, Hollingsworth AD, Kyung Yang S, Patel S, Pine DJ, Weck M. Patchy particle self-assembly via metal coordination. J Am Chem Soc. 2013;135(38):14064–7.
48. Kraft DJ, Hilhorst J, Heinen MAP, Hoogenraad MJ, Luigjes B, Kegel WK. Patchy polymer colloids with tunable anisotropy dimensions. J Phys Chem B. 2011;115(22):7175–81.
49. Désert A, Chaduc I, Fouilloux S, Taveau J-C, Lambert O, Lansalot M, Bourgeat-Lami E, Thill A, Spalla O, Ravaine S, Duguet E. High-yield preparation of polystyrene/silica clusters of controlled morphology. Polym Chem. 2012;3(5):1130.
50. Désert A, Hubert C, Fu Z, Moulet L, Majimel J, Barboteau P, Thill A, Lansalot M, Bourgeat-Lami E, Duguet E, Ravaine S. Synthesis and site-specific functionalization of tetravalent, hexavalent, and dodecavalent silica particles. Angew Chem Int Ed. 2013;52(42):11068–72.
51. Kraft DJ, Ni R, Smallenburg F, Hermes M, Yoon K, Weitz DA, van Blaaderen A, Groenewold J, Dijkstra M, Kegel WK. Surface roughness directed self-assembly of patchy particles into colloidal micelles. Proc Natl Acad Sci. 2012;109(27):10787–92.
52. Sacanna S, Irvine WTM, Chaikin PM, Pine DJ. Lock and key colloids. Nature. 2010;464(7288):575–8.
53. Mely Ramírez L, Milner ST, Snyder CE, Colby RH, Velegol D. Controlled flats on spherical polymer colloids. Langmuir. 2010;26(10):7644–9.
54. Ramírez LM, Smith AS, Unal DB, Colby RH, Velegol D. Self-assembly of doublets from flattened polymer colloids. Langmuir. 2012;28(9):4086–94.

55. Ramírez LM, Michaelis CA, Rosado JE, Pabón EK, Colby RH, Velegol D. Polloidal chains from self-assembly of flattened particles. Langmuir. 2013;29(33):10340–5.
56. Cayre O, Paunov VN, Velev OD. Fabrication of asymmetrically coated colloid particles by microcontact printing techniques. J Mater Chem. 2003;13(10):2445.
57. Jiang S, Granick S. A simple method to produce trivalent colloidal particles. Langmuir. 2009;25(16):8915–8.
58. Seidel P, Ravoo BJ. Preparation of microscale polymer janus particles by sandwich microcontact printing. Macromol Chem Phys. 2016;217(13):1467–72.
59. Gangwal S, Cayre OJ, Velev OD. Dielectrophoretic assembly of metallodielectric Janus particles in AC electric fields. Langmuir. 2008;24:13312–20.
60. Gangwal S, Cayre OJ, Bazant MZ, Velev OD. Induced-charge electrophoresis of metallodielectric particles. Phys Rev Lett. 2008.100:058302.
61. Smoukov SK, Gangwal S, Marquez M, Velev OD. Reconfigurable responsive structures assembled from magnetic Janus particles. Soft Matter. 2009;5:1285–92.
62. Gangwal S, Pawar A, Kretzschmar I, Velev OD. Programmed assembly of metallodielectric patchy particles in external AC electric fields. Soft Matter. 2010;6:1413–8.
63. Kretzschmar I, Song JH. Surface-anisotropic spherical colloids in geometric and field confinement. Curr Opin Colloid Interface Sci. 2011;16:84–95.
64. Rossi L, Sacanna S, Irvine WTM, Chaikin PM, Pine DJ, Philipse AP. Cubic crystals from cubic colloids. Soft Matter. 2011;7:4139–42.
65. Champion JA, Katare YK, Mitragotri S. Making polymeric micro- and nanoparticles of complex shapes. Proc Natl Acad Sci. 2007;104(29):11901–4.
66. Vutukuri HR, Imhof A, van Blaaderen A. Fabrication of polyhedral particles from spherical colloids and their self-assembly into rotator phases. Angew Chem Int Ed. 2014;53(50):13830–4.
67. Barry E, Dogic Z. Entropy driven self-assembly of nonamphiphilic colloidal membranes. Proc Natl Acad Sci. 2010;107(23):10348–53.
68. Qi W, de Graaf J, Qiao F, Marras S, Manna L, Dijkstra M. Ordered two-dimensional superstructures of colloidal octapod-shaped nanocrystals on flat substrates. Nano Lett. 2012;12(10):5299–303. PMID: 22938387.
69. Vutukuri HR, Smallenburg F, Badaire S, Imhof A, Dijkstra M, van Blaaderen A. An experimental and simulation study on the self-assembly of colloidal cubes in external electric fields. Soft Matter. 2014;10:9110–9.
70. Bharti B, Velev OD. Assembly of reconfigurable colloidal structures by multi-directional field-induced interactions. Langmuir. 2015;31:7897–908.
71. Ristenpart WD, Aksay IA, Saville DA. Electrically guided assembly of planar superlattices in binary colloidal suspensions. Phys Rev Lett. 2003;90:128303.
72. Khalil KS, Sagastegui A, Li Y, Tahir MA, Socolar JES, Wiley BJ, Yellen BB. Binary colloidal structures assembled through Ising interactions. Nat Commun. 2012;3:794.
73. Nych A, Ognysta U, Škarabot M, Ravnik M, Žumer S, Muševič I. Assembly and control of 3D nematic dipolar colloidal crystals. Nat Commun. 2013;4:1489.
74. Mangold K, Leiderer P, Bechinger C. Phase transitions of colloidal monolayers in periodic pinning arrays. Phys Rev Lett. 2003;90:158302.
75. Mikhael J, Roth J, Helden L, Bechinger C. Archimedean-like tiling on decagonal quasicrystalline surfaces. Nature. 2008;454:501–4.
76. Demirors AF, Pillai PP, Kowalczy B, Grzybowski M. Colloidal assembly directed by virtual magnetic moulds. Nature. 2013;503:99–103.
77. Bianchi E, Kahl G, Likos CN. Inverse patchy colloids: from microscopic description to mesoscopic coarse-graining. Soft Matter. 2011;7(18):8313.
78. Bianchi E, van Oostrum PDJ, Likos CN, Kahl G. Inverse patchy colloids: synthesis, modeling and self-organization. Curr Opin Colloid Interface Sci. 2017;30:8–15.
79. Denkov N, Velev O, Kralchevski P, Ivanov I, Yoshimura H, Nagayama K. Mechanism of formation of two-dimensional crystals from latex particles on substrates. Langmuir. 1992;8(12):3183–90.

80. Choueiri RM, Galati E, Thérien-Aubin H, Klinkova A, Larin EM, Querejeta-Fernández A, Han L, Xin HL, Gang O, Zhulina EB, Rubinstein M, Kumacheva E. Surface patterning of nanoparticles with polymer patches. Nature. 2016;538(7623):79–83.
81. Capone B, Coluzza I, LoVerso F, Likos CN, Blaak R. Telechelic star polymers as self-assembling units from the molecular to the macroscopic scale. Phys Rev Lett. 2012;109(23):238301.
82. Alward DB, Kinning DJ, Thomas EL, Fetters LJ. Effect of arm number and arm molecular weight on the solid-state morphology of poly(styrene-isoprene) star block copolymers. Macromolecules. 1986;19(1):215–24.
83. Thomas EL, Alward DB, Kinning DJ, Martin DC, Handlin DL, Fetters LJ. Ordered bicontinuous double-diamond structure of star block copolymers: a new equilibrium microdomain morphology. Macromolecules. 1986;19(8):2197–202.
84. Zhao Y, Berger R, Landfester K, Crespy D. Polymer patchy colloids with sticky patches. Polym Chem. 2014;5(2):365–71.
85. Nikoubashman A, Lee VE, Sosa C, Prud'homme RK, Priestley RD, Panagiotopoulos AZ. Directed assembly of soft colloids through rapid solvent exchange. ACS Nano. 2016;10(1):1425–33.
86. Sosa C, Liu R, Tang C, Qu F, Niu S, Bazant MZ, Prud'homme RK, Priestley RD. Soft multifaced and patchy colloids by constrained volume self-assembly. Macromolecules. 2016;49(9):3580–5.
87. Li N, Panagiotopoulos AZ, Nikoubashman A. Structured nanoparticles from the self-assembly of polymer blends through rapid solvent exchange. Langmuir. 2017;33:6021–28.
88. Higuchi T, Tajima A, Yabu H, Shimomura M. Spontaneous formation of polymer nanoparticles with inner micro-phase separation structures. Soft Matter. 2008;4:1302–5.
89. Cheng L, Zhang G, Zhu L, Chen D, Jiang M. Nanoscale tubular and sheetlike superstructures from hierarchical self-assembly of polymeric Janus particles. Angew Chem Int Ed. 2008;47(52):10171–4.
90. Gröschel AH, Walther A, Löbling TI, Schmelz J, Hanisch A, Schmalz H, Müller AHE. Facile, solution-based synthesis of soft, nanoscale janus particles with tunable janus balance. J Am Chem Soc. 2012;134(33):13850–60.
91. Hanisch A, Gröschel AH, Förtsch M, Drechsler M, Jinnai H, Ruhland TM, Schacher FH, Müller AHE. Counterion-mediated hierarchical self-assembly of an ABC Miktoarm Star terpolymer. ACS Nano. 2013;7(5):4030–41. PMID: 23544750.
92. Gröschel AH, Walther A, Löbling TI, Schacher FH, Schmalz H, Müller AHE. Guided hierarchical co-assembly of soft patchy nanoparticles. Nature. 2013;503(7475):247–51.
93. Ke Y, Ong LL, Shih WM, Yin P. Three-dimensional structures self-assembled from DNA bricks. Science. 2012;338(6111):1177–83.
94. Cademartiri L, Bishop KJM. Programmable self-assembly. Nat Mater. 2015;14(1):2–9.
95. Lawrence DS, Jiang T, Levett M. Self-assembling supramolecular complexes. Chem Rev. 1995;95(6):2229–60.
96. Ratner BD, Bryant SJ. Biomaterials: where we have been and where we are going. Annu Rev Biomed Eng. 2004;6(1):41–75.
97. Coluzza I, Muller HG, Frenkel D. Designing refoldable model molecules. Phys Rev E. 2003;68(4):046703.
98. Coluzza I. A coarse-grained approach to protein design: Learning from design to understand folding. PLoS One. 2011;6(7):e20853.
99. Coluzza I, Dellago C. The configurational space of colloidal patchy polymers with heterogeneous sequences. J Phys Condens Matter. 2012;24(28):284111.
100. Coluzza I. Computational protein design: a review. J Phys Condens Matter. 2017;29(14):143001.
101. Coluzza I, van Oostrum PDJ, Capone B, Reimhult E, Dellago C. Design and folding of colloidal patchy polymers. Soft Matter. 2013;9(3):938–44.
102. Coluzza I, van Oostrum PDJ, Capone B, Reimhult E, Dellago C. Sequence controlled self-knotting colloidal patchy polymers. Phys Rev Lett. 2013;110(7):075501.

103. Goubault C, Leal-Calderon F, Viovy JL, Bibette J. Self-assembled magnetic nanowires made irreversible by polymer bridging. Langmuir. 2005;21(9):3725–9.
104. Vutukuri HR, Demirors AF, Peng B, van Oostrum PDJ, Imhof A, van Blaaderen A. Colloidal analogues of charged and uncharged polymer chains with tunable stiffness. Angew Chem Int Ed. 2012;51(45):11249–53.
105. Biswal SL, Gast AP. Mechanics of semiflexible chains formed by poly(ethylene glycol)-linked paramagnetic particles. Phys Rev E Stat Nonlin Soft Matter Phys. 2003;68(2 Pt 1):021402.
106. Byrom J, Han P, Savory M, Biswal SL. Directing assembly of DNA-coated colloids with magnetic fields to generate rigid, semiflexible, and flexible chains. Langmuir. 2014;30(30):9045–52.
107. Bannwarth MB, Kazer SW, Ulrich S, Glasser G, Crespy D, Landfester K. Well-defined nanofibers with tunable morphology from spherical colloidal building blocks. Angew Chem Int Ed. 2013;52(38):10107–11.
108. Bannwarth MB, Utech S, Ebert S, Weitz DA, Crespy D, Landfester K. Colloidal polymers with controlled sequence and branching constructed from magnetic field assembled nanoparticles. ACS Nano. 2015;9(3):2720–8. PMID: 25695858.
109. Merrifield RB. Solid phase peptide synthesis. I. the synthesis of a tetrapeptide. J Am Chem Soc. 1963;85(14):2149–54.

Multi-Scale Approach for Self-Assembly and Protein Folding

5

Oriol Vilanova, Valentino Bianco, and Giancarlo Franzese

5.1   Introduction

Self-assembly, driven by non-covalent interactions like van der Waals and hydrogen bonds, fulfills a crucial role in the supramolecular organization and assembling of the biological matter. Living being are complex organisms where matter is self-organized on different length scales in a kind of biological network. A primary role is played by the proteins that control the majority of chemical processes in the cell. Proteins are synthesized as long polymer chains composed by hundreds of monomers, taken from 20 different amino acids. Among the huge amount of possible amino acid sequences, nature has selected those that are able to fold into specific functionalized structures, known as native protein structures. The relation between the sequence code and the native structure and the way that proteins fold represents a significant example of biological self-assembly.

A crucial aspect of the protein folding that involves the interplay with solvent is related to the hydrophobic or hydrophilic nature of the amino acids. The protein folding mechanism is dominated by the dynamics of water that drives the collapse of the hydrophobic protein core and stabilizes the tertiary protein structure [1–3]. However, many proteins exhibit a limited range of temperatures T and pressures P where they are able to maintain the native structure [4–19]. Beyond those

O. Vilanova • G. Franzese (✉)
Secció de Física Estadística i Interdisciplinària, Departament de Física de la Matèria Condensada, Facultat de Física & Institute of Nanoscience and Nanotechnology (IN2UB), Universitat de Barcelona, Martí i Franquès 1, 08028 Barcelona, Spain
e-mail: gfranzese@ub.edu

V. Bianco
Computational Physics Group, Faculty of Physics, Universität Wien, Sensengasse 8/10, 1090 Vienna, Austria

© Springer International Publishing AG, part of Springer Nature 2017
I. Coluzza (ed.), *Design of Self-Assembling Materials*,
https://doi.org/10.1007/978-3-319-71578-0_5

T- and P-ranges a protein unfolds, with a consequent loss of its tertiary structure and functionality.

At high T protein unfolding is due to the thermal fluctuations that disrupt the protein structure. Open protein conformations increase the entropy S minimizing the global Gibbs free energy $G \equiv H - TS$, where H is the total enthalpy. By decreasing T proteins can crystallize but, if the nucleation of water is avoided, some proteins denaturate [5, 7, 12, 14, 20–23]. Usually such phenomena are observed below the melting line of water, although in some cases cold denaturation occurs above the 0°C, as in the case of the yeast frataxin [14].

Cold- and P-denaturation of proteins have been related to the equilibrium properties of hydration water [24–34]. However, the interpretation of this mechanism is still largely debated [35–47].

Protein denaturation is observed also upon pressurization [4, 6, 13, 19, 35]. A possible explanation of the high-P denaturation is the loss of internal cavities, sometimes present in the folded states of proteins [45]. Denaturation at negative P has been experimentally observed [48] and simulated [34, 48, 49] recently. Pressure denaturation is usually observed at $100\,\text{MPa} \lesssim P \lesssim 600\,\text{MPa}$, and rarely at higher P, unless the tertiary structure is engineered with stronger covalent bonds [11].

5.2   Hawley Theory

In 1971, Hawley proposed a theory [50] explaining the stability region (SR) for proteins as a close region in the T–P plane Fig. 5.1. The SR represents the region where a protein folds into its native conformation. Such a simple two-state theory is based on the assumption that the folding (f) unfolding (u) transition is a first order phase transition and that equilibrium thermodynamics holds during the denaturation, neglecting all the details about the protein structure. Following [50] we can express the free energy difference $\Delta G \equiv G_f - G_u$ between the free energy of the folded (G_f) and the unfolded (G_u) states as a quadratic function of T and P

$$\Delta G(P, T) = \frac{\Delta \beta}{2}(P - P_0)^2 + 2\Delta\alpha(P - P_0)(T - T_0) + \tag{5.1}$$

$$-\frac{\Delta C_P}{2T_0}(T - T_0)^2 + \Delta V_0(P - P_0) - \Delta S_0(T - T_0) + \Delta G_0$$

where T_0 and P_0 are the temperature and pressure of the reference state point (ambient conditions); ΔV and ΔS are the volume and entropy variation upon unfolding respectively; $\alpha \equiv (\partial V/\partial T) = -(\partial S/\partial P)_T$ is the thermal expansivity factor, related to the isobaric thermal expansion coefficient α_P by $\alpha_P = \alpha/V$; $C_P \equiv T(\partial S/\partial T)_P$ is the isobaric heat capacity; $\beta \equiv (\partial V/\partial P)_T$ is the isothermal compressibility factor related to the isothermal compressibility K_T by the relation $K_T = (\beta/V)$ and ΔG_0 is an integration constant. Equation (5.1) represents an

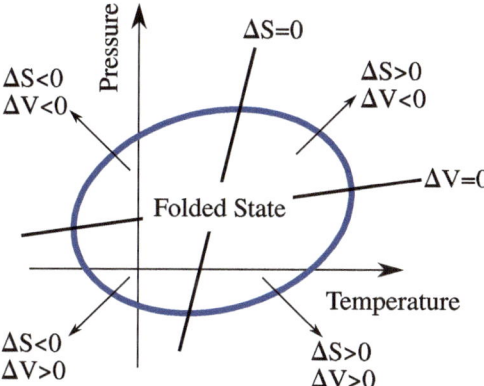

Fig. 5.1 Stability region of proteins according to the Hawley theory [50]. A protein is stable in its native state for temperature and pressures inside the elliptic curve. Across the folding (f) unfolding (u) transition the system undergoes positive or negative variations of the total (protein plus the solvent) volume ΔV and total entropy ΔS according to the figure. Straight lines join the points where the $f \longrightarrow u$ transition is isoentropic or isochoric, and where the tangent to the SR is horizontal or vertical, respectively

ellipsis given the constraint

$$\Delta \alpha^2 > \Delta C_P \Delta \beta / T_0 \qquad (5.2)$$

which is guaranteed by the different sign of ΔC_P and $\Delta \beta$ as reported by Hawley [50]. Although the calculation of Hawley is based on a Taylor expansion of the free energy variation truncated to the second order, adding more terms in Eq. (5.1) results in minor corrections that do not affect the close elliptic-like coexistence curve. All in all, the Hawley model is a phenomenological theory that makes strong assumptions on the $f \longrightarrow u$ process [51]. Nevertheless, its ability to describe all the denaturation mechanisms actually observed in experiments makes it a good test for models of protein unfolding.

5.3 A Coarse-Grain Model for Solvated Protein

5.3.1 Bulk Water Model

We present a coarse-grain model for protein water interactions based on a lattice representation of the protein, embedded in explicit water. The adopted water model is "many-body" [34, 42, 52–59].

The coarse-grain representation of the many-body interactions is based on a discretization of the available molecular volume V into a fixed number N_0 of cells, each with volume $v \equiv V/N_0 \geq v_0$, where v_0 is the water excluded volume. Each cell accommodates at most one molecule with the average O–O distance between next neighbor water molecules given by $r = v^{1/3}$. To each cell we associate a variable $n_i = 1$ if $v_0/v > 0.5$, and $n_i = 0$ otherwise. Hence, n_i is a discretized density field replacing the water translational degrees of freedom. The Hamiltonian of bulk water is

$$\mathcal{H} \equiv \sum_{ij} U(r_{ij}) - JN_{\mathrm{HB}}^{(\mathrm{b})} - J_\sigma N_{\mathrm{coop}}. \tag{5.3}$$

The first term accounts for the van der Waals interaction and is modeled with a Lennard-Jones potential

$$\sum_{ij} U(r_{ij}) \equiv 4\epsilon \sum_{ij} \left[\left(\frac{r_0}{r_{ij}} \right)^{12} - \left(\frac{r_0}{r_{ij}} \right)^6 \right] \tag{5.4}$$

where the sum runs over all the water molecules i and j at O–O distance r_{ij} and $\epsilon \equiv 5.8\,\mathrm{kJ/mol}$. We assume $U(r) \equiv \infty$ for $r < r_0 \equiv v_0^{1/3} = 2.9\,\text{Å}$ that is the water molecule hard core (water van der Waals diameter). Moreover, we apply a cutoff to the potential for $r > r_c \equiv 6r_0$.

The second term represents the directional and covalent components of the hydrogen bond (HB), where

$$N_{\mathrm{HB}}^{(\mathrm{b})} \equiv \sum_{\langle ij \rangle} n_i n_j \delta_{\sigma_{ij},\sigma_{ji}} \tag{5.5}$$

is the number of bulk HBs and the sum runs over the neighboring cells. $\sigma_{ij} = 1, \ldots, q$ is the bonding index of molecule i with respect to the neighbor molecule j. $\delta_{ab} = 1$ if $a = b$, 0 otherwise. Each water molecule can form up to four HBs. Each HB is stable if the hydrogen atom H is in a range of $[-30°; 30°]$ with respect to the O–O axes. Hence, only 1/6 of the entire range of values $[0, 360°]$ for the $\widehat{\mathrm{OOH}}$ angle is associated to a bonded state, leading to the choice $q = 6$ to account correctly for the entropy variation due to HB formation and breaking. For each HB the energy decreases an amount $-J$, where $J/4\epsilon = 0.3$. According to [52], a good choice for the parameters is $\epsilon = 5.5\,\mathrm{kJ/mol}$, $J/4\epsilon = 0.5$ and $J_\sigma/4\epsilon = 0.05$. For such a choice, the average HB energy is $\sim 23\,\mathrm{kJ/mol}$. Here, to account for the ions in solution that are always present in the cellular environment where natural proteins are embedded, we decrease the ratio J/J_σ, that modify the bulk phase diagram in a qualitative way similar to that induced by ions [60]. For such a choice we find that the HB energy is $\sim 20\,\mathrm{kJ/mol}$.

The third term in the Hamiltonian represents the cooperative interaction between the HBs. Such an effect is due to the quantum many-body interaction [61]: the formation of a new HB affects the electron distribution around the molecule favoring the formation of the following HB in a local tetrahedral structure [62]. To mimic the cooperativity of the HBs we introduce an effective interaction between the bonding indexes of a molecule

$$N_{coop} \equiv \sum_i n_i \sum_{(l,k)_i} \delta_{\sigma_{ik},\sigma_{il}} \tag{5.6}$$

where $(l,k)_i$ indicates each of the six different pairs of the four indices σ_{ij} of a molecule i. The choice $J_\sigma/4\epsilon \equiv 0.05 \ll J$ guarantees the asymmetry between the two HB terms.

The formation of HBs leads to an open network of molecules, giving rise to a lower density state. We include this effect into the model assuming that, for each HB formed, the volume V increases by $v_{HB}^{(b)}/v_0 = 0.5$, corresponding to the average volume increase between high-density ices VI and VIII and low-density (tetrahedral) ice Ih. We assume that the HBs do not affect the distance r between first neighbor molecules, consistent with experiments [62]. Hence, the HB formation does not affect the $U(r)$ term.

The total bulk volume $V^{(b)}$ is

$$V^{(b)} \equiv Nv_0 + N_{HB}^{(b)} v_{HB}^{(b)}. \tag{5.7}$$

5.3.2 Modeling Protein-Water Interplay

The protein is modeled as a hydrophobic self-avoiding lattice polymer and it is embedded into the cell partition of the system. Despite its simplicity, lattice protein models are still widely used in the contest of protein folding [25, 26, 32, 34, 63, 64] because of their versatility and the possibility to better understand many mechanisms of the protein dynamics. Each protein residue (polymer bead) occupies one cell, without affecting its volume. In the present study, we do not consider the presence of cavities into the protein structure.

To simplify the discussion, we assume that no residue–residue interactions occur and that the residue–water interaction vanishes. This implies that the protein has several ground states, all with the same maximum number n_{max} of residue–residue contacts. Our results hold also when such interactions are restored [34]. Here we adopt the symbol Φ to refer to hydrophobic residues. The protein interface affects the water–water properties in the hydration shell, here we define it as the layer of first neighbor water molecules in contact with the protein. There are many numerical and experimental evidences supporting the hypothesis that water–water HBs in the hydration shell are more stable and more correlated with respect to bulk HBs [65–69]. We account for this by replacing J of Eq. (5.3) with $J_\Phi > J$ for water–water HBs at the Φ interface. This choice, according to Muller [70], ensures the water enthalpy compensation upon cold-denaturation [33].

In addition to the stronger/stabler water–water HBs in the Φ shell, we incorporate into the model also the larger density fluctuations at the Φ interface with respect to the bulk, observed in water hydrating Φ solutes [40, 67]. Such an increase of density fluctuations results in a Φ hydration shell that, at ambient pressure, is more compressible than bulk water. Although it is still matter of debate whether the average density of water at the Φ interface is larger or smaller with respect to the average bulk water density [71–75], there are evidences showing that such density fluctuations reduce upon pressurization [40, 67, 76, 77]. Hence, if we attribute this P-dependence of the Φ shell density to the interfacial HB properties, we can assume that the average volume associated to HBs formed in the Φ shell is

$$v_{HB}^{(\Phi)}/v_{HB,0}^{(\Phi)} \equiv 1 - k_1 P \qquad (5.8)$$

where $v_{HB,0}^{(\Phi)}$ is the volume change associated to the HB formation in the Φ hydration shell at $P = 0$ and k_1 is a positive factor. According to [34], we could add other polynomial terms to Eq. (5.8), although such terms would not affect our results as long as $P < 1/k_1$ [78].

The total volume V, including the contributions coming from the HBs formed in the Φ shell is

$$V \equiv N v_0 + N_{HB}^{(b)} v_{HB}^{(b)} + N_{HB}^{(\Phi)} v_{HB}^{(\Phi)}. \qquad (5.9)$$

where $N_{HB}^{(\Phi)}$ is the number of HBs in the Φ shell.

In the following we fix $k_1 = 1 v_0/4\epsilon$, $v_{HB,0}^{(\Phi)}/v_0 = v_{HB}^{(b)}/v_0 = 0.5$ and $J_\Phi/J = 1.83$. Our findings are robust with respect to a change of parameters.

5.3.3 Simulation Details

We study proteins with 30 residues using Monte Carlo simulations in the isobaric-isothermal ensemble, i.e. at constant P, constant T and constant number of particles. All the simulations start with a protein in a completely folded conformation. The water bonding indexes σ are equilibrated using a cluster algorithm. The protein is equilibrated using corner flips, pivot and crankshaft moves [79] Fig. 5.2. Along the simulation we calculate the average number of residue–residue contacts to estimate the protein compactness. For each state point we sample $\sim 10^4$ independent protein conformations.

5.3.4 Results

We assume that the protein is folded if the average number of residue–residue contacts is $n_{rr} \geq 50\% \, n_{max}$. In Fig. 5.3 we show the calculated SR for the protein, consistent with the Hawley theory [9, 50]. We observe that the SR has an

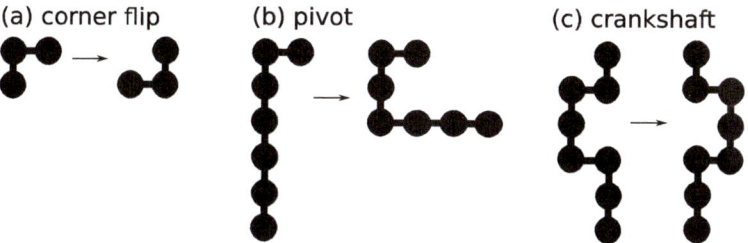

Fig. 5.2 Possible protein moves. (**a**) In the *corner flip* move a monomer in a corner configuration can jump to the opposite corner. (**b**) In the *pivot* move we randomly choose one monomer, acting as a pivot, and rotate the shorter side chain, with respect to it. (**c**) The *crankshaft* move consists in choosing at random an axis passing through two monomers, and rotating the included residues

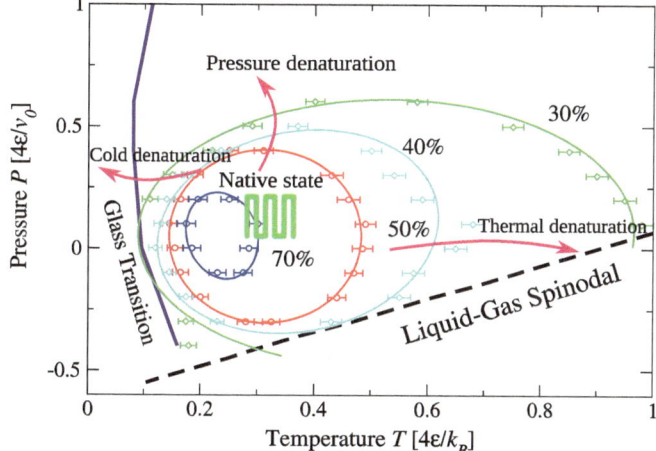

Fig. 5.3 $P - T$ stability region of the protein, calculated with Monte Carlo simulations. The symbols mark the state points where the protein has the same average residue–residue contact's number $n_{rr}/n_{max} = 30\%$, 40%, 50% and 70%, corresponding to different percentage of compactness. Elliptic lines are guides for the eyes. The "glass transition" line defines the temperatures below which the system does not equilibrate. The spinodal line marks the stability limit of the liquid phase at high P with respect to the gas at low P; k_B is the Boltzmann constant

elliptic shape that is preserved independently of the compactness we adopt as the reference for the folded state, underlying that the folded⟶unfolded transition is a continuous process. Proteins undergo heat-, cold-, and P-unfolding. The folded protein minimizes the number of hydrated Φ residues, reducing the energy cost of the interface, as expected.

Upon increasing T at constant P, we observe that the model reproduces the expected *entropy-driven* unfolding. The entropy S increases both for the opening of the protein and for the larger decrease of water–water HBs.

Decreasing T at constant P leads to open protein conformations that minimize the Gibbs free energy. The difference in energy gain between bulk HBs and HBs at the Φ interface results in a competing mechanism. A bulk water molecule can form up to four HBs, while the water molecules at the Φ interface can form up to three HBs, although stronger. Hence, reducing the exposed protein surface maximizes the possible number of bulk HBs, while increasing the hydrated protein surface maximizes the interfacial HBs. At low T the number of HBs in the Φ shell saturates and the only way for the system to further minimize the internal energy is by increasing $N_{HB}^{(\Phi)}$, i.e. by unfolding the protein. Hence, the cold denaturation is an *energy-driven* process.

Upon an isothermal increase of P the protein denaturates. We find that this change is associated to a decrease of $N_{HB}^{(b)}$ and an increase of $N_{HB}^{(\Phi)}$ leading to a net decrease of V at high P, as a consequence of the compressible Φ shell, Eq. (5.8). At high P, the PV term of the Gibbs free energy dominates the $f \longrightarrow u$ process. Hence, the water contribution to the high-P denaturation induces a *density-driven* process, resulting in a denser Φ shell with respect to the bulk.

Finally, lowering P toward negative values results in a negative contribution $(Pv_{HB}^{(\Phi)} - J_\Phi)N_{HB}^{(\Phi)}$, leading to a decrease in enthalpy when the protein opens up and $N_{HB}^{(\Phi)}$ increases. Therefore we find that at negative P the denaturation process is *enthalpy-driven*.

By varying the parameters $v_{HB}^{(\Phi)}$ and J_Φ we find that the first is relevant for the P-denaturation, as we expected because it dominates the volume contribution to the Gibbs free energy, while the second affects the stability range in T. Both effects combine in a non-trivial way to regulate the SR, shifting, shrinking, and dilating the SR, although the elliptic shape is preserved [78].

We can estimate also the entropy change and volume change, respectively indicated with ΔS and ΔV, during the $f \longleftrightarrow u$ process. First, we calculate the average volume of the unfolded—completely stretched—protein V_u and of the folded—maximum compactness—protein V_f in a wide range of T and P, equilibrating water around the fixed protein conformations. From the difference $\Delta V \equiv V_f - V_u$ we calculate ΔS using the Clapeyron relation $dP/dT = \Delta S/\Delta V$ along the SR curve [50]. The Clapeyron equation holds, in principle, only along first order phase transitions. Though the $f \longleftrightarrow u$ process is not necessary a phase transitions, the calculation of ΔS and ΔV represents a good model test to compare our results with the Hawley's theory [50]. In Fig. 5.4 we show that our findings match with the theoretical predictions: T-denaturation is accompanied by a positive entropy variation $\Delta S > 0$ at high T and an entropic penalty $\Delta S < 0$ at low T; P-denaturation is accompanied by a decrease of volume $\Delta V < 0$ at high P and an increase of volume $\Delta V > 0$ at low P. In particular, at $P = 0.3(4\epsilon/v_0)$, corresponding to ≈ 500 MPa, we find that $\Delta V \approx -2.5v_0$, hence $|P\Delta V| = 0.75(4\epsilon) \approx 17$ kJ/mol, very close to the typical reported value of 15 kJ/mol [13].

Therefore our coarse-grain model allows to understand how water contributes to the temperature- and pressure-denaturation of proteins. Accounting for stronger and more stable HBs in the hydrophobic hydration shell with respect to the bulk and for a more compressible hydrophobic hydration shell our model reproduces a close stability region for proteins with the expected elliptic-like shape in the T–P plane,

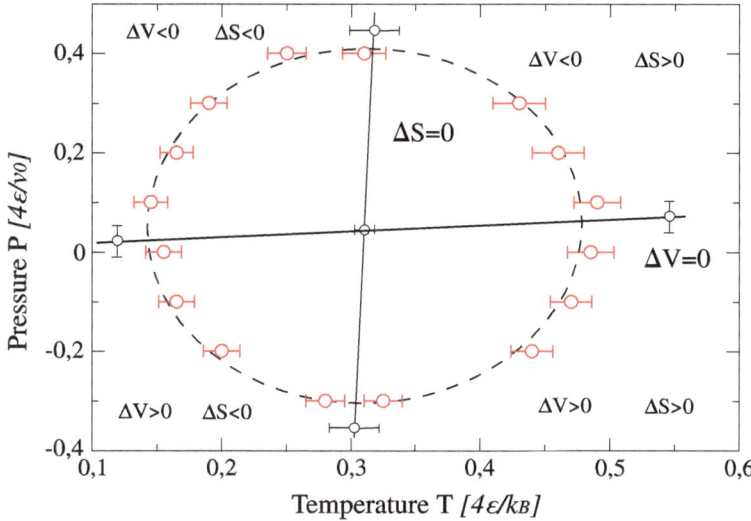

Fig. 5.4 Volume change ΔV and entropy change ΔS for the $f \longrightarrow u$ process in the T–P plane. Solid lines connect state points with isochoric $\Delta V = 0$ and isoentropic $\Delta S = 0$ denaturation. Red points mark the SR, adapted from Fig. 5.3. The loci $\Delta V = 0$ and $\Delta S = 0$ have a positive slope and intersect the SR at the turning points with $dT/dP|_{SR} = 0$ and $dT/dP|_{SR} = \infty$, respectively

consistent with theory [50]. We find that cold denaturation is energy-driven, while unfolding at high pressures and negative pressures are density- and enthalpy-driven by water, respectively.

5.4 Protein Adsorption onto NPs: Kinetics of the Protein-Corona Formation

Nanoparticles (NPs) are small scale objects with dimensions in the range from 1 nm (10^{-9} m) up to 100 nm. The main feature of these particles relies on the fact that some of their properties differ completely from the bulk material they are made of. Furthermore, their properties strongly depend on the particle size. Due to their small size, the surface to volume ratio of NPs is much higher than that of macroscopic objects. This feature, combined with the high surface free energy of NPs, confers NPs a high level of chemical reactivity.

The NPs particular interactions with biological systems make them very promising tools for medical applications and could allow to simultaneously perform therapeutics and diagnostics (*theranostics*) [80–82]. In particular, experiments show that NPs are able to cross cellular barriers, including the strongest defense we have in our body, the blood-brain barrier. Therefore, the fact that NPs interact directly with the biological machinery [83] represents an opportunity to deliver drugs to specific targets hidden in the most inaccessible spots within the cells for treating illnesses that challenge us, such as cancer or neurodegenerative diseases [84].

In the last years, industry has started to produce NPs at a large scale, with an increasing rate, due to their industrial, commercial, and medical applications. Thousands of commercial products, such as textiles, cosmetics, or paints, contain NPs, and nanomaterials are nowadays used in medical treatments, electronics, or food. In order to keep producing and using them at the industrial scale safely, many experimental studies [85] of the nanotoxicology impact [86] of specific NPs have started to evaluate the hazard of exposing the environment, living beings and humans to them [87–90].

Yet, very little is known about the mechanisms regulating the interactions of NPs with biological systems [91]. Acquiring such knowledge would allow us to predict if the interaction of such small-scale materials with living organisms would potentially be dangerous before even performing elaborated experiments.

It has been confirmed that NPs absorb proteins and other biomolecules from the environment forming a complex that is known as the "protein corona" (Fig. 5.5) [92–97]. Certain proteins are also able to prevent the adsorption of other molecules, or to modify the chemical properties of the NP surface, and even to determine the path and final localization of the NP in a living organism [81, 98].

Material surfaces exposed to biological environments are commonly modified by the adsorption of biomolecules, such as proteins and lipids, already present

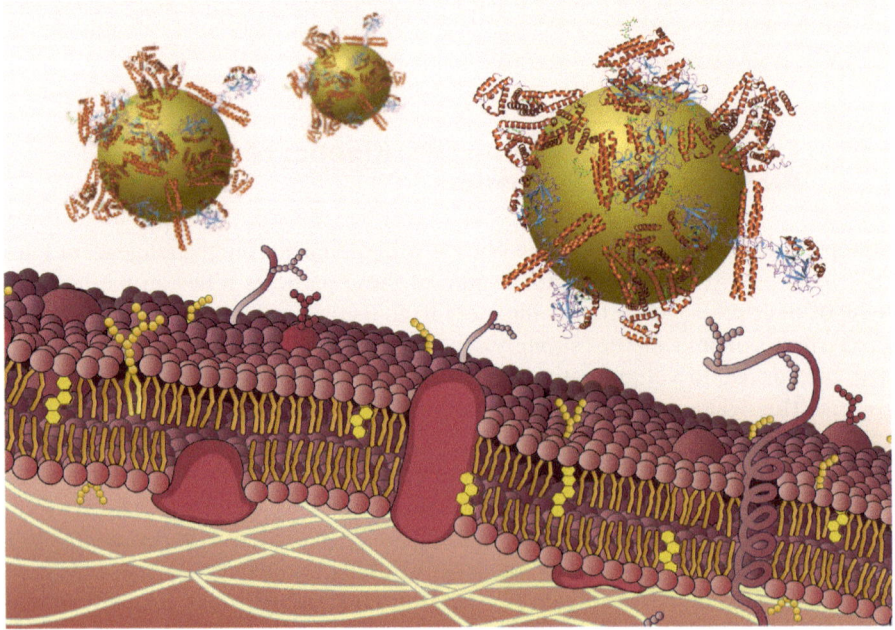

Fig. 5.5 Pictorial representation of the NP-"Protein corona" complex near a cellular membrane

in solution. It has been shown that the cellular response to any material in a biological medium is mediated by the adsorbed biomolecular layer, rather than the bare material itself [93,94]. For this reason, the scientific community has focused its attention to the fact that NP interactions with living organisms must be also mediated by the adsorbed protein layer [99, 100]. It is now understood that the biological identity of the NP is characterized by the distribution of the distinct adsorbed proteins on its surface [101–104]. This collection of adsorbed biomolecules is in fact dynamic and evolves in time [93–95, 97, 100]. The effective particle made by the protein-NP-corona is essential to understand how NPs interact with cells in biological media [99]. Our aim is to understand and explain how the interactions between biological macromolecules and NPs in solution lead to the formation of the effective particle and to its evolution over time scales that are relevant in the biological context.

In the following we address the establishment of a theoretical framework and the determination of the key features that could give a simple, but complete, picture of protein–NP interactions. We focus on parameters that describe the NP surface at the molecular level, such as curvature, charge or surface chemistry, that determine which kinds of proteins end up binding from a complex biological fluid, e.g., blood plasma or cell media [103].

The adoption of a multi-scale approach in this respect is particularly suitable because it provides the theoretical framework to study a variety of aspects that occur at different length- and time-scales, ranging from the atomistic scale ($\sim$0.1 nm and $\sim$1 ns) to the mesoscale ($\sim$1 μm and $\sim$100 s). We, therefore, first establish phenomenological models for protein–protein and protein–NP interactions, in solvent [105]. Next, we validate these models with, two types of NPs using available experimental data [106]. Finally we perform predictions, based on our theoretical models, and we test them by comparing with experimental data. In particular, experimentally checking the kinetics and the protein corona composition allows us to assess the predictive power of the models when we use only our knowledge on physico-chemical properties of the nanomaterial and the environmental conditions [107].

5.5 Methodology

5.5.1 Multi-Scale Modeling

For the purpose of developing this research we make use of state-of-the-art techniques of computational and theoretical modeling. Thereby, we adopt a multi-scale approach to carry out this study. Multi-scale modeling is based on the idea that each specific problem is characterized by multiple scales of time and space. There is a need to consider every scale to understand the whole picture, given the relevance of different mechanisms occurring at each scale. This approach is particularly appropriate for the case of biological systems at the nanoscale, such as the interactions of NPs with macromolecules in presence of water. In this situation, the length scales range from angstroms to micrometers while the time scales span more than ten orders of magnitude.

We separate our analysis into different levels of description, based on the observation that each length scale has an associated time scale. By these means, each level of description is related to a range of length and time scales that partially overlap with the consecutive levels. To this end, each level is characterized by specific phenomena, which can be useful to understand features of other, maybe more complex, phenomena at higher levels of description. In our case, we consider the following levels of description: (1) the molecular and macromolecular level, ranging from 1 Å to 100 nm and from 10 fs to 1 μs , and (2) the mesoscale level, ranging from 1 nm to 1 μm and from 10 ns to 1 s. The next step is to consider, for example, the problem of NP aggregation or the interaction of NPs with cellular membranes, that would span a range of length scales up to 100 μm and time-scales up to hours.

5.5.2 Theoretical Framework

We make use of suitable computational and theoretical techniques to approach each level of description. The smaller scales are characterized by the explicit consideration of water molecules in the solvent, as mediators of the processes that occur at the macromolecular scale. However, dealing with explicit solvent models hampers the efficiency of computation in a dramatic way. This is why we adopt the coarse-grain model of water introduced in the previous sections [52, 55, 58] to describe the macromolecule–solvent interactions [34]. This is a convenient strategy that allows for wider and more efficient options to explore these systems. The model can be efficiently simulated using the Monte Carlo (MC) method [52, 55, 108, 109] and can also be treated analytically, allowing us to reach a deeper understanding of the fundamental mechanisms [110–112].

At the larger scales, where we focus our attention on the interaction of NPs with solutions containing a large number of proteins, we use an implicit solvent approach. In this case, we take into account the effects of the solvent by means of modified dynamics and effective interactions. Therefore, we also consider the proteins and the NPs as coarse-grain objects. We adopt a Molecular Dynamics (MD) simulation scheme using Langevin dynamics (LD), that allows for the determination of the kinetic properties of the system. Moreover, we introduce a framework that describes protein–protein interactions via effective potentials, and we make use of the well-established DLVO theory for colloidal dispersions in solution to describe the protein–NP interactions [107].

Finally, we develop a Non-Langmuir Dynamical Rate Equation (NLDRE) model which is phenomenological analytical theory to describe the protein adsorption kinetics [107]. This theory provides us tools to extrapolate the simulation results of protein adsorption obtained with MD, to much longer time-scales. With this approach we are able to predict kinetic processes over experimentally relevant time-scales of the order of several minutes, far beyond the limits of the computational capacities [107].

5.5.3 Experimental Validation

The collaborative work of theorist and experimentalists is a key factor for the successful development of theoretical models with predictive capability. The experimental data is fundamental for the parametrization of the theoretical models, yet it is crucial for the validation and verification process of the theoretical predictions. To this goal we focus on a simplified version of a multicomponent protein solution as it would be the human blood plasma. In particular, we adopt a "model plasma" containing only three proteins that are representative of the extremely large number of protein forming the human blood [107]. This step is essential to design experiments suitable for comparison with simulations.

We follow a workflow where we can first measure in a controlled experimental setup all the data that are relevant for defining the phenomenological parameters of the theoretical model. In this way, the preliminary experimental results serve to calibrate the theoretical model.

Next we perform our simulations and analytic prediction based on these phenomenological input parameters. We make our calculation under conditions that can be reproduced in experiments, in simple cases with NPs in bicomponent protein solutions, for a direct comparison. Then we design a set of simple experiments to validate the simulation results obtained under identical conditions.

Finally, to test the predictive power of the theoretical tools, we consider more complex situations, such as NPs in three-component solutions made of proteins that are competing for the NP surface. Under these conditions the experiments reveal a *memory effect* of the protein corona, not predicted by our initial model. As a consequence, we modify the model in a way that allows us to reproduce the effect. Although this result exposes a limit of our initial approach, it also led us to propose a mechanism to rationalize the memory effect. Specifically, we show that to account for the memory effect it is necessary to model the self-assembly of the proteins on the NPs including irreversible structural changes [107].

5.5.4 Simulation Details and Results

We develop a high-performance suite of simulation codes able to run on Graphical Processing Units (GPUs) [113]. We use a molecular dynamics (MD) approach to simulate the interactions of NPs with protein solutions. We consider a simulation box with periodic boundary conditions containing one single NP fixed to the center of the box. We separate the simulation box into two regions. The inner region contains the NP, while we use the external region as a reservoir to control the protein concentration inside the inner region [107]. We treat the solvent in an implicit way, by introducing a coarse-grain protein–protein and protein–NP interaction model using continuous shouldered potentials [105, 114, 115]. We also adopt a Langevin dynamics (LD) integration scheme to take into account the collisions with the solvent molecules for the correct diffusion of the proteins.

We follow a sequential protocol to insert different kinds of proteins at selected times during a simulation. For example, in Fig. 5.6 we show three snapshots of a simulation containing a 100 nm diameter silica (SiO_2) NP at 0.1 mg/ml in a protein solution of Human Serum Albumin (HSA), Transferrin (Tf), and Fibrinogen (Fib). The first snapshot shows the equilibrium state after introducing HSA at 0.07 mg/ml and Tf at 0.07 mg/ml to the system, at the same moment that we introduce Fib at 0.005 mg/ml. The second snapshot is a transient state, where Fib begins adsorbing to the NP surface. In the third snapshot we show how the high affinity of Fib allows it to displace HSA and Tf proteins from the surface, while the system has not yet reached equilibrium because this is a kinetically slow process (Fig. 5.7a).

In Fig. 5.7 we show the effect of increasing the concentration of the proteins during the incubation step to the adsorption kinetics of the third protein. In panel (a) the NP is incubated in a solution of HSA and Tf at a low concentration of 0.07 mg/ml each, and we introduce Fib at 0.005 mg/ml which is a protein with a much higher surface affinity. We make use of our NLDRE to fit the parameters to the simulation data and we extrapolate the adsorption kinetics to much longer time-scales. In panel (b) we show the adsorption kinetics of a system where we initially incubate the NP in a solution of HSA and Tf at a much higher concentration of 3.5 mg/ml each. Using the analytical extrapolation to the simulation data we find that the Fib adsorption kinetics are at least one order of magnitude slower compared to the previous situation.

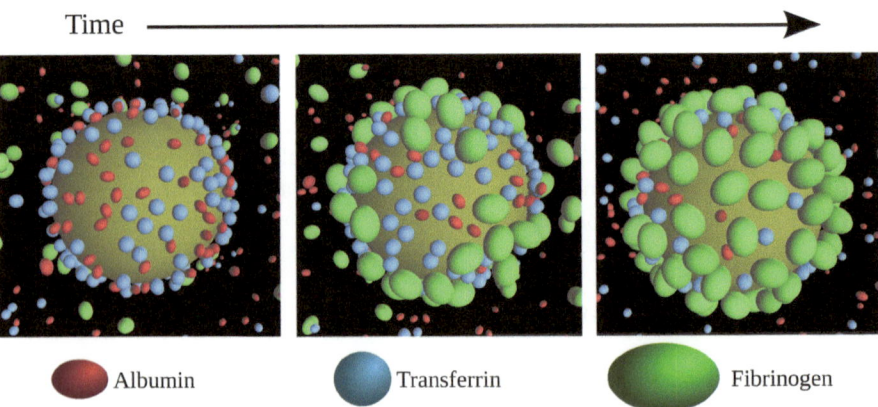

Time

Albumin   Transferrin   Fibrinogen

Fig. 5.6 Simulation snapshots at three different times for a system containing one silica NP of 100 nm diameter in a solution containing Human Serum Albumin (HSA), Transferrin (Tf) and Fibrinogen (Fib). The introduction of each protein is done sequentially in three steps: HSA first, Tf second, and Fib third

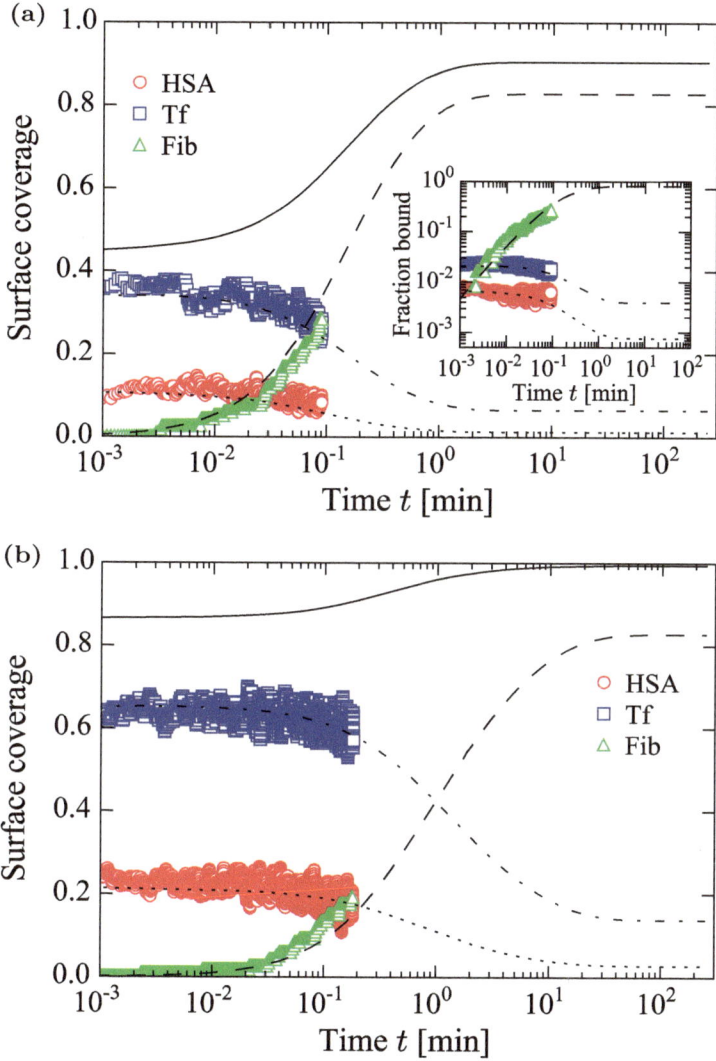

Fig. 5.7 Simulation results (symbols) and analytical extrapolation (lines) using NLDRE of the protein corona kinetics in a system containing 100 nm silica NPs at 0.1 mg/ml in a protein solution containing HSA (red circles), Tf (blue squares), and Fib (green triangles) at different concentrations. The NPs are initially incubated in a solution containing HSA and Tf, with the subsequent introduction of Fib at 0.005 mg/ml. We also show the analytical extrapolation using the NLDRE theory to the simulation data for each protein (dashed lines), and the total surface coverage (solid lines). (**a**) NP surface coverage after incubation in a solution of HSA at 0.07 mg/ml and Tf at 0.07 mg/ml, inset: Fraction bound of proteins shown for comparison with the surface coverage. (**b**) NP surface coverage after incubation in a solution of HSA at 3.5 mg/ml and Tf at 3.5 mg/ml

5.6   Conclusions

We combine theory with experiments to introduce a new framework consisting in experimentally-supported computational simulations and theoretical models as a methodology to obtain reliable predictive results for protein folding and protein-NP self-assembly.

We establish a multi-scale modeling approach to deal with the different phenomenologies characteristic to each level of description. Based on atomistic Molecular Dynamics simulations over nanoseconds up to microseconds time-scales, we define a coarse-grain water model that allows us to simulate hydrated systems [116] at length- and time-scales that are relevant for biological processes. This model allows us to study by Monte Carlo simulations protein structural changes and folding-unfolding events [34]. We recently extended this approach to study protein design [117].

With this in mind, we introduce a set of computational tools (BUBBLES [113]) that allows us to study phenomenological protein–protein and protein–NP interactions [105] with Langevin Dynamics simulations up to the time scale of seconds. Next we introduce a phenomenological theory based on rate equation (NLDRE) with which we can extrapolate our simulation results to time scales of hours, allowing us direct comparison with the experimental results.

In particular, we first study the protein corona formation for a simple system of polystyrene NPs in a solution containing only a single kind of proteins (Transferrin). We introduce many-body interactions to explain the multilayer adsorption mechanism, the protein corona kinetics, and the soft-corona/hard-corona characterization [106].

Finally, we study more complicated systems by introducing NPs in solutions made of multiple types of proteins, choosing a set of proteins that compete for the NP surface. This allows for the emergence of competitive protein adsorption and assembly on top of the NP, a rich and complex playground that we exploit to discover and understand new and unexpected features [107].

Acknowledgements We are thankful to M. Bernabei, C. Calero, L. E. Coronas, F. Leoni, N. Pagès, and A. Zantop for helpful discussions. O.V. and G.F. acknowledge the support of Spanish MINECO grant FIS2012-31025 and FIS2015-66879-C2-2-P. I. C. acknowledges the support from the Austrian Science Fund (FWF) Grant No. 26253-N27. V.B. acknowledges the support of the FWF Grant No. 2150-N36 and P 26253-N27.

References

1. Levy Y, Onuchic JN. Water and proteins: a love-hate relationship. Proc Natl Acad Sci USA. 2004;101(10):3325–6.
2. Levy Y, Onuchic JN. Mechanisms of protein assembly: lessons from minimalist models. Acc Chem Res. 2006;39(2):135–42.

3. Raschke TM. Water structure and interactions with protein surfaces. Curr Opin Struct Biol. 2006;16(2):152–9.
4. Zipp A, Kauzmann W. Pressure denaturation of metmyoglobin. Biochemistry 1973;12(21):4217–28.
5. Privalov PL. Cold denaturation of proteins. Crit Rev Biochem Mol Biol. 1990;25(4):281–305.
6. Hummer G, Garde S, García AE, Paulaitis ME, Pratt LR. The pressure dependence of hydrophobic interactions is consistent with the observed pressure denaturation of proteins. Proc Natl Acad Sci. 1998;95(4):1552–5.
7. Meersman F, Smeller L, Heremans K. Pressure-assisted cold unfolding of proteins and its effects on the conformational stability compared to pressure and heat unfolding. High Pressure Res. 2000;19(1–6):263–8.
8. Lassalle MW, Yamada H, Akasaka K. The pressure-temperature free energy-landscape of staphylococcal nuclease monitored by (1)H NMR. J Mol Biol. 2000;298(2):293–302.
9. Smeller L. Pressure-temperature phase diagrams of biomolecules. Biochim Biophys Acta Protein Struct Mol Enzymol. 2002;1595(1–2):11–29.
10. Herberhold H, Winter R. Temperature- and pressure-induced unfolding and refolding of ubiquitin: a static and kinetic Fourier transform infrared spectroscopy study. Biochemistry 2002;41(7):2396–401.
11. Lesch H, Stadlbauer H, Friedrich J, Vanderkooi JM. Stability diagram and unfolding of a modified cytochrome c: what happens in the transformation regime? Biophys J. 2002;82(3):1644–53.
12. Ravindra R, Winter R. On the temperature–pressure free-energy landscape of proteins. Chem Phys Chem. 2003;4(4):359–65.
13. Meersman F, Dobson CM, Heremans K. Protein unfolding, amyloid fibril formation and configurational energy landscapes under high pressure conditions. Chem Soc Rev. 2006;35(10):908–17.
14. Pastore A, Martin SR, Politou A, Kondapalli KC, Stemmler T, Temussi PA. Unbiased cold denaturation: low- and high-temperature unfolding of yeast Frataxin under physiological conditions. J Am Chem Soc. 2007;129(17):5374–5.
15. Wiedersich J, Köhler S, Skerra A, Friedrich J. Temperature and pressure dependence of protein stability: the engineered fluorescein-binding lipocalin FluA shows an elliptic phase diagram. Proc Natl Acad Sci USA. 2008;105(15):5756–61.
16. Maeno A, Matsuo H, Akasaka K. The pressure–temperature phase diagram of hen lysozyme at low pH. Biophysics 2009;5:1–9.
17. Somkuti J, Mártonfalvi Z, Kellermayer MSZ, Smeller L. Different pressure–temperature behavior of the structured and unstructured regions of titin. Biochim. Biophys. Acta 2013;1834(1):112–8.
18. Somkuti J, Jain S, Ramachandran S, Smeller L. Folding-unfolding transitions of Rv3221c on the pressure-temperature plane. High Pressure Res. 2013;33(2):250–7.
19. Nucci NV, Fuglestad B, Athanasoula EA, Wand AJ. Role of cavities and hydration in the pressure unfolding of T4 lysozyme. Proc Natl Acad Sci USA. 2014;111(38):13846–51.
20. Griko YV, Privalov PL, Sturtevant JM, Venyaminov SY. Cold denaturation of staphylococcal nuclease. Proc Natl Acad Sci. 1988;85(10):3343–7.
21. Goossens K, Smeller L, Frank J, Heremans K. Pressure-tuning the conformation of bovine pancreatic trypsin inhibitor studied by fourier-transform infrared spectroscopy. Eur J Biochem. 1996;236(1):254–62.
22. Nash DP, Jonas J. Structure of the pressure-assisted cold denatured state of ubiquitin. Biochem Biophys Res Commun. 1997;238(2):289–91.
23. Nash DP, Jonas J . Structure of pressure-assisted cold denatured lysozyme and comparison with lysozyme folding intermediatesă. Biochemistry 1997;36(47):14375–83.
24. De Los Rios P, Caldarelli G. Putting proteins back into water. Phys Rev E. 2000;62(6):8449–52.

25. Marqués MI, Borreguero JM, Stanley HE, Dokholyan NV. Possible mechanism for cold denaturation of proteins at high pressure. Phys Rev Lett. 2003;91(13):138103.
26. Patel BA, Debenedetti PG, Stillinger FH, Rossky PJ. A water-explicit lattice model of heat-, cold-, and pressure-induced protein unfolding. Biophys J. 2007;93(12):4116–27.
27. Athawale MV, Goel G, Ghosh T, Truskett TM, Garde S. Effects of lengthscales and attractions on the collapse of hydrophobic polymers in water. Proc Natl Acad Sci USA. 2007;104(3):733–8.
28. Nettels D, Müller-Späth S, Küster F, Hofmann H, Haenni D, Rüegger S, Reymond L, Hoffmann A, Kubelka J, Heinz B, Gast K, Best RB, Schuler B. Single-molecule spectroscopy of the temperature-induced collapse of unfolded proteins. Proc Natl Acad Sci. 2009;106(49):20740–5.
29. Best RB, Mittal J. Protein simulations with an optimized water model: cooperative helix formation and temperature-induced unfolded state collapse. J Phys Chem B. 2010;114(46):14916–23.
30. Jamadagni SN, Bosoy C, Garde S. Designing heteropolymers to fold into unique structures via water-mediated interactions. J Phys Chem B. 2010;114(42):13282–8.
31. Badasyan AV, Tonoyan SA, Mamasakhlisov YS, Giacometti A, Benight AS, Morozov VF. Competition for hydrogen-bond formation in the helix-coil transition and protein folding. Phys Rev E Stat Nonlin Soft Matter Phys. 2011;83(5 Pt 1):051903.
32. Matysiak S, Debenedetti PG, Rossky PJ. Role of hydrophobic hydration in protein stability: a 3D water-explicit protein model exhibiting cold and heat denaturation. J Phys Chem B. 2012;116(28):8095–104.
33. Bianco V, Iskrov S, Franzese G. Understanding the role of hydrogen bonds in water dynamics and protein stability. J Biol Phys. 2012;38(1):27–48.
34. Bianco V, Franzese G. Contribution of water to pressure and cold denaturation of proteins. Phys Rev Lett. 2015;115(10):108101.
35. Paschek D, García AE. Reversible temperature and pressure denaturation of a protein fragment: a replica exchange molecular dynamics simulation study. Phys Rev Lett. 2004;93(23):238105.
36. Paschek D, Gnanakaran S, Garcia AE. Simulations of the pressure and temperature unfolding of an alpha-helical peptide. Proc Natl Acad Sci USA. 2005;102(19):6765–70.
37. Sumi T, Sekino H. Possible mechanism underlying high-pressure unfolding of proteins: formation of a short-period high-density hydration shell. Phys Chem Chem Phys. 2011;13(35):15829–32.
38. Coluzza I. A coarse-grained approach to protein design: learning from design to understand folding. PloS One 2011;6(7):e20853.
39. Dias CL. Unifying microscopic mechanism for pressure and cold denaturations of proteins. Phys Rev Lett. 2012;109(4):048104.
40. Das P, Matysiak S. Direct characterization of hydrophobic hydration during cold and pressure denaturation. J Phys Chem B. 2012;116(18):5342–8.
41. Sarma R, Paul S. Effect of pressure on the solution structure and hydrogen bond properties of aqueous N-methylacetamide. Chem Phys. 2012;407:115–23.
42. Franzese G, Bianco V. Water at biological and inorganic interfaces. Food Biophys. 2013;8(3):153–69.
43. Abeln S, Vendruscolo M, Dobson CM, Frenkel D. A simple lattice model that captures protein folding, aggregation and amyloid formation. PloS One 2014;9(1):e85185.
44. Yang C, Jang S, Pak Y. A fully atomistic computer simulation study of cold denaturation of a β-hairpin. Nat Commun. 2014;5:5773.
45. Roche J, Caro JA, Norberto DR, Barthe P, Roumestand C, Schlessman JL, Garcia AE, García-Moreno BE, Royer CA, Garc\'ia AE, Garcia-Moreno BE, Royer CA. Cavities determine the pressure unfolding of proteins. Proc Natl Acad Sci USA. 2012;109(18):6945–50.
46. Nisius L, Grzesiek S. Key stabilizing elements of protein structure identified through pressure and temperature perturbation of its hydrogen bond network. Nat Chem. 2012;4(9):711–7.

47. van Dijk E, Varilly P, Knowles TP, Frenkel D, Abeln S. Consistent treatment of hydrophobicity in protein lattice models accounts for cold denaturation. arXiv e-prints 2015;116(7):078101.
48. Larios E. Gruebele M. Protein stability at negative pressure. Methods (San Diego, Calif.) 2010;52(1):51–6.
49. Hatch HW, Stillinger FH, Debenedetti PG. Computational study of the stability of the miniprotein Trp-cage, the GB1 β-hairpin, and the AK16 peptide, under negative pressure. J Phys Chem B. 2014;118(28):7761–9.
50. Hawley SA. Reversible pressure–temperature denaturation of chymotrypsinogen. Biochemistry 1971;10(13):2436–42.
51. Meersman F, Smeller L, Heremans K. Protein stability and dynamics in the pressure-temperature plane. Biochim Biophys Acta. 2006;1764(3):346–54.
52. Stokely K, Mazza MG, Stanley HE, Franzese G. Effect of hydrogen bond cooperativity on the behavior of water. Proc Natl Acad Sci USA. 2010;107:1301–6.
53. Strekalova EG, Mazza MG, Stanley HE, Franzese G. Large decrease of fluctuations for supercooled water in hydrophobic nanoconfinement. Phys Rev Lett. 2011;106:145701.
54. Franzese G, Bianco V, Iskrov S. Water at interface with proteins. Food Biophys. 2011;6:186–98. https://doi.org/10.1007/s11483-010-9198-4.
55. Mazza MG, Stokely K, Pagnotta SE, Bruni F, Stanley HE, Franzese G. More than one dynamic crossover in protein hydration water. Proc Natl Acad Sci. 2011;108(50):19873–8.
56. Bianco V, Vilanova O, Franzese G. Polyamorphism and polymorphism of a confined water monolayer: liquid–liquid critical point, liquid–crystal and crystal–crystal phase transitions. In: Proceedings of perspectives and challenges in statistical physics and complex systems for the next decade: a conference in honor of Eugene Stanley and Liacir Lucen; 2013. p. 126–49.
57. de los Santos F, Franzese G. Understanding diffusion and density anomaly in a coarse-grained model for water confined between hydrophobic walls. J Phys Chem B. 2011;115:14311–20.
58. Bianco V, Franzese G. Critical behavior of a water monolayer under hydrophobic confinement. Sci Rep. 2014;4:4440.
59. Coronas LE, Bianco V, Zantop A, Franzese G. Liquid–liquid critical point in 3D many-body water model. arXiv e-prints, October 2016.
60. Corradini D, Gallo P. Liquid–liquid coexistence in NaCl aqueous solutions: a simulation study of concentration effects. J Phys Chem B. 2011;115(48):14161–6.
61. Hernández de la Peña L, Kusalik PG. Temperature dependence of quantum effects in liquid water. J Am Chem Soc. 2005;127(14):5246–51.
62. Soper AK, Antonietta Ricci M. Structures of high-density and low-density water. Phys Rev Lett. 2000;84(13):2881–4.
63. Lau FK, Dill KA. A lattice statistical mechanics model of the conformational and sequence spaces of proteins. Macromolecules 1989;22(10):3986–97.
64. Caldarelli G, De Los Rios P. Cold and warm denaturation of proteins. J Biol Phys. 2001;27(2–3):229–41.
65. Dias CL, Ala-Nissila T, Karttunen M, Vattulainen I, Grant M. Microscopic mechanism for cold denaturation. Phys Rev Lett. 2008;100(11):118101–4.
66. Petersen CL, Tielrooij K-J, Bakker HJ. Strong temperature dependence of water reorientation in hydrophobic hydration shells. J Chem Phys. 2009;130(21):214511.
67. Sarupria S. Garde S. Quantifying water density fluctuations and compressibility of hydration shells of hydrophobic solutes and proteins. Phys Rev Lett. 2009;103(3):37803.
68. Tarasevich YI. State and structure of water in vicinity of hydrophobic surfaces. Colloid J. 2011;73(2):257–66.
69. Davis JG, Gierszal KP, Wang P, Ben-Amotz D. Water structural transformation at molecular hydrophobic interfaces. Nature 2012;491(7425):582–5.
70. Muller N. Search for a realistic view of hydrophobic effects. Acc Chem Res. 1990;23(1):23–8.
71. Lum K, Chandler D, Weeks JD. Hydrophobicity at small and large length scales. J Phys Chem B. 1999;103(22):4570–7.

72. Schwendel D, Hayashi T, Dahint R, Pertsin A, Grunze M, Steitz R, Schreiber F. Interaction of water with self-assembled monolayers: neutron reflectivity measurements of the water density in the interface region. Langmuir 2003;19(6):2284–93.

73. Jensen TR, Østergaard Jensen M, Reitzel N, Balashev K, Peters GH, Kjaer K, Bjørnholm T. Water in contact with extended hydrophobic surfaces: direct evidence of weak dewetting. Phys Rev Lett. 2003;90(8):86101.

74. Doshi DA, Watkins EB, Israelachvili JN, Majewski J. Reduced water density at hydrophobic surfaces: effect of dissolved gases. Proc Natl Acad Sci USA. 2005;102(27):9458–62.

75. Godawat R, Jamadagni SN, Garde S. Characterizing hydrophobicity of interfaces by using cavity formation, solute binding, and water correlations. Proc Natl Acad Sci USA. 2009;106(36):15119–24.

76. Ghosh T, García AE, Garde S. Molecular dynamics simulations of pressure effects on hydrophobic interactions. J Am Chem Soc. 2001;123(44):10997–1003.

77. Dias CL, Chan HS. Pressure-dependent properties of elementary hydrophobic interactions: ramifications for activation properties of protein folding. J Phys Chem B. 2014;118(27):7488–509.

78. Bianco V, Pagès Gelabert N, Coluzza I, Franzese G. How the stability of a folded protein depends on interfacial water properties and residue–residue interactions. arXiv e-prints, April 2017.

79. Frenkel D, Smit B. Understand molecular simulations. San Diego/London: Academic; 2002.

80. Habash M, Reid G. Microbial biofilms: their development and significance for medical device-related infections. J Clin Pharmacol. 1999;39(9):887–98.

81. Salvati A, Pitek AS, Monopoli MP, Prapainop K, Bombelli FB, Hristov DR, Kelly PM, Åberg C, Mahon E, Dawson KA. Transferrin-functionalized nanoparticles lose their targeting capabilities when a biomolecule corona adsorbs on the surface. Nat Nanotechnol. 2013;8(2):137–43.

82. Ding H-M, Ma YQ. Design strategy of surface decoration for efficient delivery of nanoparticles by computer simulation. Sci Rep. 2016;6:26783.

83. De Simone A, Spadaccini R, Temussi PA, Fraternali F. Toward the understanding of MNEI sweetness from hydration map surfaces. Biophys J. 2006;90(9):3052–61.

84. Puntes VF. Design and pharmacokinetical aspects for the use of inorganic nanoparticles in radiomedicine. Br J Radiol. 2016;89(1057):20150210.

85. Lindman S, Lynch I, Thulin E, Nilsson H, Dawson KA, Linse S. Systematic investigation of the thermodynamics of HSA adsorption to N-iso-propylacrylamide/N-tert-butylacrylamide copolymer nanoparticles. Effects of particle size and hydrophobicity. Nano Lett. 2007;7(4):914–20.

86. Dawson KA, Salvati A, Lynch I. Nanotoxicology: nanoparticles reconstruct lipids. Nat Nano. 2009;4(2):84–5.

87. Lundqvist M, Stigler J, Elia G, Lynch I, Cedervall T, Dawson KA. Nanoparticle size and surface properties determine the protein corona with possible implications for biological impacts. Proc Natl Acad Sci. 2008;105(38):14265–70.

88. Pratap N, Casey A, Lynch I. Tenuta T, Dawson KA. Preparation, characterization and ecotoxicological evaluation of four environmentally relevant species of n- isopropylacrylamide and n-isopropylacrylamide-co-n-tert-butylacrylamide copolymer nanoparticles. Aquat Toxicol. 2009;92:146–54.

89. Rivera Gil P, Oberdörster G, Elder A, Puntes VF, Parak WJ. Correlating physico-chemical with toxicological properties of nanoparticles: the present and the future. ACS Nano. 2010;4(10):5227–31.

90. Corbo C, Molinaro R, Parodi A, Furman NET, Salvatore F, Tasciotti E. The impact of nanoparticle protein corona on cytotoxicity, immunotoxicity and target drug delivery. Nanomedicine. 2016;11(1):81–100.

91. Lynch I, Cedervall T, Lundqvist M, Cabaleiro-Lago C, Linse S, Dawson KA. The nanoparticle-protein complex as a biological entity; a complex fluids and surface science challenge for the 21st century. Adv Colloid Interface Sci. 2007;134–135:167–74.

92. Cedervall T, Lynch I, Lindman S, Berggård T, Thulin E, Nilsson H, Dawson KA, Linse S. Understanding the nanoparticle-protein corona using methods to quantify exchange rates and affinities of proteins for nanoparticles. Proc Natl Acad Sci USA. 2007;104(7):2050–5.
93. Lynch I, Salvati A, Dawson KA. Protein-nanoparticle interactions: what does the cell see? Nat Nanotechnol. 2009;4(9):546–7.
94. Walczyk D, Bombelli FB, Monopoli MP, Lynch I, Dawson KA. What the cell "sees" in bionanoscience. J Am Chem Soc. 2010;132(16):5761–8.
95. Casals E, Pfaller T, Duschl A, Oostingh GJ, Puntes VF. Time evolution of the nanoparticle protein corona. ACS Nano. 2010;4(7):3623–32.
96. Dell'Orco D, Lundqvist M, Oslakovic C, Cedervall T, Linse S. Modeling the time evolution of the nanoparticle-protein corona in a body fluid. PLoS One 2010;5(6):e10949.
97. Milani S, Bombelli FB, Pitek AS, Dawson KA, Rädler J, Baldelli Bombelli F. Reversible versus irreversible binding of transferrin to polystyrene nanoparticles: soft and hard corona. ACS Nano. 2012;6(3):2532–41.
98. Pitek AS, O'Connell D, Mahon E, Monopoli MP, Bombelli FB, Dawson KA. Transferrin coated nanoparticles: study of the bionano interface in human plasma. PLoS One. 2012;7(7):e40685.
99. Monopoli MP, Åberg C, Salvati A, Dawson KA. Biomolecular coronas provide the biological identity of nanosized materials. Nat Nanotechnol. 2012;7(12):779–86.
100. Lundqvist M, Stigler J, Cedervall T, Berggård T, Flanagan MB, Lynch I, Elia G, Dawson K. The evolution of the protein corona around nanoparticles: a test study. ACS Nano. 2011;5(9):7503–9.
101. Shapero K, Fenaroli F, Lynch I, Cottell DC, Salvati A, Dawson KA. Time and space resolved uptake study of silica nanoparticles by human cells. Mol BioSyst. 2011;7:371–8.
102. Salvati A, Åberg C, dos Santos T, Varela J, Pinto P, Lynch I, Dawson KA. Experimental and theoretical comparison of intracellular import of polymeric nanoparticles and small molecules: toward models of uptake kinetics. Nanomed Nanotechnol Biol Med. 2011;7(6):818–26.
103. Monopoli MP, Walczyk D, Campbell A, Elia G, Lynch I, Bombelli FB, Dawson KA. Physical-chemical aspects of protein corona: relevance to in vitro and in vivo biological impacts of nanoparticles. J Am Chem Soc. 2011;133(8):2525–34.
104. Monopoli MP, Bombelli FB, Dawson KA. Nanobiotechnology: nanoparticle coronas take shape. Nat Nano. 2011;6(1):11–2
105. Vilaseca P, Dawson KA, Franzese G. Understanding and modulating the competitive surface-adsorption of proteins through coarse-grained molecular dynamics simulations. Soft Matter. 2013;9(29):6978–85.
106. Vilanova O. Bionanointeractions: interactions between nanoscopic systems and biological macromolecules in solution. PhD thesis, Universitat de Barcelona. 2018.
107. Vilanova O, Mittag JJ, Kelly PM, Milani S, Dawson KA, Rädler JO, Franzese G. Understanding the kinetics of protein–nanoparticle corona formation. ACS Nano. 2016;10(12):10842–50
108. Kumar P, Franzese G, Stanley HE. Dynamics and thermodynamics of water. J Phys Condens Matter. 2008;20(24):244114.
109. Mazza MG , Stokely K, Strekalova EG, Stanley HE, Franzese G. Cluster Monte Carlo and numerical mean field analysis for the water liquid–liquid phase transition. Comput Phys Commun. 2009;180(4):497–502.
110. Franzese G, Malescio G, Skibinsky A, Buldyrev SV, Stanley HE. Metastable liquid-liquid phase transition in a single-component system with only one crystal phase and no density anomaly. Phys Rev E. 2002;66(5):51206.
111. Franzese G, Stanley HE. A theory for discriminating the mechanism responsible for the water density anomaly. Physica A. 2002;314(1–4):508–13.
112. Franzese G, Stanley HE. Liquid–liquid critical point in a Hamiltonian model for water: analytic solution. J Phys Condens Matter. 2002;14(9):2201–9.
113. https://github.com/bubbles-suite/BUBBLES (2015).

114. Franzese G. Differences between discontinuous and continuous soft-core attractive potentials: the appearance of density anomaly. J Mol Liq. 2007;136(3):267–73.
115. Vilaseca P, Franzese G. Isotropic soft-core potentials with two characteristic length scales and anomalous behaviour. J Non-Cryst Solids. 2011;357(2):419–26.
116. Vilanova O, Franzese G. Structural and dynamical properties of nanoconfined supercooled water. arXiv.org, arXiv:1102.2864. 2011.
117. Bianco V, Franzese G, Dellago C, Coluzza I. Role of water in the selection of stable proteins at ambient and extreme thermodynamic conditions. Phys Rev X. 2017;7:021047.

Index

© Springer International Publishing AG, part of Springer Nature 2017
I. Coluzza (ed.), *Design of Self-Assembling Materials*,
https://doi.org/10.1007/978-3-319-71578-0

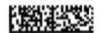